VIEWING STONES
OF NORTH AMERICA

VIEWING STONES
OF NORTH AMERICA

A CONTEMPORARY PERSPECTIVE

by Thomas S. Elias

with contributions by Paul Gilbert, Rick Stiles, and Richard Turner

Floating World Editions
in cooperation with
VSANA
Viewing Stone Association
of North America

First edition 2014
Published by Floating World Editions, Inc.
26 Jack Corner Road, Warren, CT 06777

In cooperation with the
Viewing Stone Association of North America

Printed in China
ISBN 978-1-891640-72-8
Library of Congress Cataloging-in-Publication data available

Jacket front, *Ancient Sentinel*, see Gallery 12;
frontispiece, *Cascade*, see Gallery 34;
jacket back, *The Dance*, see Gallery 118.

CONTENTS

Dedicated to my friend
and ardent stone collector,
Ralph Johnson

PREFACE

My interest in viewing stone appreciation began and developed in China and Japan. I was fortunate to be able to travel and work in these countries frequently from 1978 to the present. I have made over thirty trips to China and nearly as many separate trips to Japan. The majority of these trips included visits to stone exhibitions and festivals, stone markets, private stone collections, classical and modern botanical gardens, and bookshops in both countries throughout the years.

My first encounter with stones that were collected and used as viewing stones happened in the spring of 1978 when along with nine other scientists I was invited by the Chinese Academy of Sciences to tour and evaluate university and botanical garden research facilities in the People's Republic of China. During our extensive tour, we visited several well-known classical Chinese gardens in Yangzhou and Suzhou. I also observed the beautiful Ying stones that were displayed in the garden section of the Imperial Palace in the Forbidden City in Beijing. The extensive use of Taihu, Taihu-like, Lingbi, and Ying stones, particularly as landscape elements, made a profound impression upon me. Since that time, I have been a student of Chinese gardens, the use of stones in these gardens, and their influence on gardens outside of China. This opened the door to learning about the use of smaller stones indoors as objects of appreciation. In the last fifteen years, my knowledge of Chinese stones has grown steadily along with the expansion of a collection of Chinese stones and a library of Chinese language stone-related books and journals. My appointment as Honorary Vice Chairman of the View Stone Association of China and as Vice President of the Guangdong Viewing

Stone Association helped me gain access to stone specialists, private collections, and other sources of information about stone appreciation.

With my wife, Hiromi, I travel to Japan often to study Japanese-style stone appreciation. These visits allow us to meet with leaders of the Nippon Suiseki Association and visit private stone collectors and dealers' collections. We have assembled a large collection of Japanese stones and Japanese-language books on the subject and have even displayed four of our Japanese stones in national exhibitions in Tokyo. These visits help us understand Asian concepts, and develop and refine our own opinions and collections. Our library of Asian-language books, journals, and other publications on stone appreciation may be one of the largest in North America. These volumes are complemented with English and other books in Western languages, all on aspects of stone appreciation and Asian gardens.

My first North American viewing stone was presented to me by William Merritt in 1994. At that time, Merritt was a member of the board of directors of the National Bonsai Foundation. I still have this small mountain-shaped piece of granite mounted in a wooden base. At that time, I was serving as director of the U.S. National Arboretum in Washington, D.C. The National Bonsai and Penjing Museum is a part of the arboretum. Soon after we completed the addition of a Chinese and an International Pavilion, we received a gift of a collection of North American stones for the museum. The stone collection grew as new Japanese, Chinese, African, and European stones were added, eventually to become the largest viewing stone collection held in a public garden or arboretum in North America. While in Washington,

I participated in the meetings and exhibits of the Potomac Stone Group and continued to refine my understanding of stones.

In 2009, we moved to southern California to devote more time to studying, writing, and lecturing about stone appreciation, Asian gardens, bonsai, and penjing. Participation in the California Aiseki-kai club meetings gave us a better appreciation of California stones. Eventually, we formed our own stone club and then joined with another club to form the Southern Breeze Tree and Stone Society. This society holds monthly meetings at the Huntington Library, Art Collections, and Botanical Garden, and stages occasional exhibits of trees and stones.

My interest in stones led to the publication of over twenty articles in magazine and newsletters along with our book *Chrysanthemum Stones: The Story of Stone Flowers* (2010). In February, 2012, I launched our web site on Chinese stone appreciation, www.vsana.org. The monthly updates posted are now viewed by a continually growing number of people in over eighty countries. In 2014 I coauthored the large format book *Spirit Stones* with noted stone authority Kemin Hu. This volume featured the artistic photographs of award winning photographer Jonathan Singer. Hiromi started our Facebook/vsana.org in the fall of 2013, and updates this site on a daily basis.

Over the years I have witnessed a growing interest in North American stone appreciation and the formation of several new stone clubs in different regions of the continent. While many adherents follow Japanese or Chinese guidelines for evaluating stones and displaying them, an increasing number are interested in establishing an independent North American context for stone appreciation. This led to the idea of publishing this volume and giving the many collectors an opportunity to display the beautiful array of natural viewing stones available on the North American continent. The enthusiastic response to this project has far exceeded my expectations, and the result is a book that I hope all can enjoy.

—Thomas S. Elias
Claremont, California

ACKNOWLEDGMENTS

My sincere thanks to the sixty-three people who took the time to prepare and submit photographs of their stones to this project. This book would not be possible without their enthusiastic support. Unfortunately, due to the large number of high quality stones submitted, not everyone had a stone selected. This was due to a space limitations and no reflection on the quality of their stones.

Several people played a major role in this project. First and foremost is my wife, Hiromi, whose patience, support, and wise counsel sustain my efforts. Rick Stiles, Richard Turner, and Paul Gilbert engaged in discussions, shared information, and served as sounding boards in this effort. Each contributed an essay in support of developing a North American framework for stone appreciation. The four panelists who helped me judge the 330 entries—Rick Stiles, Richard Turner, Glen Reusch, and Joe Grande—helped establish the criteria for determining the qualities of exceptional stones. Their patience and persistence is greatly appreciated. Each of these people played an important role in this book.

My sincere thanks to the professional photographers who contributed their photographic skills to supply the numerous higher quality images reproduced here. They include: Michael J. Colella in Maryland; Bernard Gastrich, New York; Scott Dressel-Martin, Colorado; Danielle Cuvilliner, Washington; and Jason Villamil in New Jersey.

A driving force in my interest in stone appreciation came from one of the early collectors of North American stones, Ralph Johnson. His love of stones and minerals is infectious. Ralph is highly skilled in motivating others and he has been responsible, directly and indirectly, for fostering both North American and Asian style stone appreciation cultures. His encouragement has helped me refine my knowledge and appreciation of natural stones.

A special thanks is due to my publisher, Ray Furse, for his encouragement and belief in this project, and for his help in producing this book. Liz Travato, once again, demonstrated her fine design and layout skills in making this an attractive and easy to use volume. Stephen Elias, a member of the English faculty at California Polytechnic University at Pomona reviewed the essays and made key suggestions for improving organization and flow. Mary Taylor Jensen, copy editor, also served as a reviewer of the text and helped improve the manuscript. This team played the final critical role in the production process.

Finally, I wish to thank all the people in China and Japan who have taken time to try and teach me about their respective stone appreciation cultures during our many trips to their countries. Their warmth, patience, and generous sharing of information, books, stones, and access to important private collections have shaped my understanding of Asian stone appreciation. Any mistakes or omissions are entirely my own.

CONTRIBUTORS

This book would not have been possible without the enthusiastic cooperation of the following individuals and organizations who submitted images and permitted them to be reproduced.

American Viewing Stone Center
Tony Ankowicz
Buzz Barry
Lindsay Bebb
Gudrun Benz
Ralph Bischof
Dawn Blankfield
Robert Blankfield
Peter Bloomer
Jerry Braswell
Chris Cochrane
Michael J. Colella
Donald Dupras
Sam Edge
Thomas S. Elias
Joseph Gaytan
Paul Gilbert
Bill Hertneky
Theresa Hertneky
Allan Hills
Jean Horton
Sean Horton
Larry Jackel
Ralph Johnson
Rick Klauber
Edd Kuehn
Chung Kruger
Don Kruger
Dien Liang
Alex Loughry
Brian McCarthy
Robert McKenzie

Cynthia McLeod
Ken McLeod
Gary McWilliams
Patrick Metiva
William B. Meran
Nan Morgan
Mas Nakajima
Hiromi Nakaoji
National Bonsai and Penjing Museum,
    U.S. National Arboretum
Al Nelson
Hanne Povlsen
Larry Ragle
Nina Ragle
Glenn Reusch
Racie Rhyne
Janet Roth
Martin Schmalenberg
Paul Schmidt, Jr.
Joel Schwarz
Sharon Somerfeld
Mimi Stiles
Rick Stiles
Jack Sustic
Richard Turner
William N. Valavanis
Freeman Wang
Darrell Whitley
Brent Wilson

INTRODUCTION

The purpose of this book is to acquaint the public with the diverse range of beautiful natural stones of North America that may be collected, displayed, and appreciated for their aesthetic qualities, and to help further a distinctively North American framework for stone appreciation.

North America geographically includes Canada, Greenland, the United States, Mexico, seven Central American nations, and numerous Caribbean islands; it is a vast region encompassing nearly 9.4 million square miles (24.35 million square kilometers), and ranges from arctic to tropical climate zones. The geology is diverse and complex, and thus has produced an amazing array of beautiful stones that we are only beginning to identify. As such, we should not try to judge North American stones using precepts developed for those of Asia. Japan is an island nation of 152,411 square miles, with a limited range of geological diversity and habitats; it lacks both deserts and extremely cold regions. China, on the other hand, is similar in size to the U.S., and has many comparable geological features. However, the culture of Chinese stone appreciation rose from philosophical and religious beliefs substantially different from our own.

Appreciation of viewing stones in North America is a relatively new activity that is distinct from a long-standing interest in gems and minerals. The waves of European immigrants who settled in Canada and the United States adopted for the most part ideas and attitudes about gem and mineral collecting from Europe, where one of the earliest compilations of the properties of stones, *De Mineralibus*, was published by Albertus Magnus over eight hundred years ago, in 1192. However, a similar interest did not develop in North America until long after European colonization. According to Wendell E. Wilson (1994), the natural resources of the New World were viewed purely in economic terms; the idea of appreciating the aesthetic qualities of minerals did not occur until specimens were imported from Europe. Thus it was not until the late 1700s that collecting minerals and gems was a firmly established activity in the New World, and many well-known early Americans were collectors, including President Thomas Jefferson. The activity has continued to grow in popularity over the centuries, until today possibly the largest gem and mineral show in the entire world is held each February in Tucson, Arizona.

An interest in stones, as opposed to gems and minerals, as objects to be appreciated for their natural beauty has been strongly influenced by imports of East Asian examples, and a stone appreciation culture based upon East Asian aesthetic principles has slowly developed in North America over the last century. Before that time, placing a small- to medium-sized stone on a base and viewing it as evocative of some aspect of nature was not customary in the West. Our limited vocabulary of stone appreciation has also reached the New World largely from East Asian cultures, and collectors have relied upon Japanese and Chinese criteria for acquiring, evaluating, and displaying stones in this manner.

"Viewing stones" is a generic term for unusual natural rocks that are collected, displayed, and appreciated indoors for their aesthetic qualities and for feelings they evoke. In China the term may include larger rocks for outdoor display, particularly in gardens or courtyard settings. Viewing stones displayed indoors are medium- to small-sized stones, typically mounted in carved wooden bases, ceramic or metal trays, or displayed on cushions or fabric. Size can vary from tiny hand-held stones to some weighing several tons. When displayed in a skillfully carved base or handmade tray, the result can be a work of art—one created through the marriage of a found object with a man-made counterpart and whose impact is more powerful than either of its constituent parts.

This volume is divided into two parts. The first contains essays presenting a brief background of Chinese and Japanese stone appreciation and how they have influenced North American concepts. Paul Gilbert examines the role of stones in American Indian cultures to determine the influence they may have had on our emerging stone appreciation culture. Avid stone collector and art connoisseur Rick Stiles provides a straightforward argument for establishing a new context for North American stone appreciation. Richard Turner presents a cogent case for considering a natural stone placed on a base for display as an art object, similar to the position stones held in the art and culture of imperial China. In part two, 151 outstanding stones are presented, selected from a total of 330 submitted for this project from sixty-three private and institutional collections. This is preceded by brief notes on the evaluation and selection process used to determine which stones would be included in the gallery of exceptional North American viewing stones.

Use of the term "viewing stone" throughout this book we hope will avoid any confusion with terms borrowed from Asia whose meanings may have changed in Western usage. The Japanese word suiseki, for example, is defined as a single stone that has the power to suggest something in nature far greater than itself (Matsuura, 2010). Hideo Marushima, author of the *History of Japanese Stones* (1992), describes suiseki as the art of enjoying the shape, surface patterns, and colors of a small stone as well as the imaginary world that the stone suggests. Contrary to popular belief, Marushima states that the term originated from *suiban-seki* (a stone displayed in a shallow tray filled with water), from its use in the Japanese bonsai community in the twentieth century. He maintains that some Japanese never use the word suiseki to describe a stone displayed on a base rather than in water, preferring the term *kansho-seki*, or "viewing stone," for those.

Unfortunately the word suiseki has been misunderstood and misused by some Western stone enthusiasts to refer to any type of stone. Likewise, the Chinese word *gongshi* should refer only to a select group of stones (spirit or offering stones) and not to the broad array of stones appreciated by modern Chinese collectors. In fact, gongshi is not commonly used in China today in the Chinese stone appreciation communities.

In summary, the East Asian influence has established a solid foundation for a viewing stone culture in the West and deserves due acknowledgment. However, after what has been a lengthy period of incubation many are looking to break away from traditional Chinese and Japanese guidelines and to create a new context and aesthetics—one more closely aligned to the geology and cultural history of North America. This is the first book to feature the finest North American stones from private and institutional collections. It is hoped that it may serve as a benchmark for the current state of an evolving North American stone appreciation culture.

STONE APPRECIATION EAST AND WEST

Thomas S. Elias

East Asian Heritage

The scholars and bureaucrats of Imperial China can collectively be considered the literati, as all studied classical texts needed to pass state examinations. All held philosophical and religious views shaped by Taoism, Confucianism, and Buddhism. The world was considered a place of order, with correspondences and influences between the realms of man, nature, and spirits. These beliefs engendered a respect for significant geological forms and formations, a reverence for nature, and a desire to live in harmony with it. They inspired a keen interest in unusual and fantastic-shaped stones, which were considered to encapsulate and exemplify the natural world and became objects of admiration. Individual stones in revered mountain ranges were identified and given names. At Huangshan (Yellow Mountains), for example, forty-two different distinct rock formations were identified and named in the *Yellow Mountain Stone Catalogue* (*Huangshan Shipu*) published in 1697.

The origins of stone collecting in China are shrouded in the mists of time. As early as four to five millennia ago people in the present-day Nanjing region collected beautiful agates, opals, and crystals known as rain flower pebbles from gravel pits along the Yangtze River and interred them alongside bodies. Jia Xiang Yun, in his 2010 reference *History of Chinese Stone Appreciation Culture*, points to the Zhou (1046–256 BCE) and Qin (221–206 BCE) dynasties as the formative period of Chinese stone appreciation.

At some point, perhaps as early as the Han dynasty (206 BCE–220 CE), large fantastic-shaped stones were brought into courtyards of homes and incorporated into gardens. The smooth, water-worn stones from Lake Tai fascinated the literati. Workers broke these stones loose and transported them, often via the Grand Canal, to estates and gardens of wealthy scholar-officials, and later, to those of a growing merchant class. Bizarrely shaped stones found at or near the surface in Lingbi County in northern Anhui Province were harvested and transported at great cost.

Smaller stones were brought indoors to be displayed or to adorn the desks of scholars, hence the name scholars' rocks. A writer, poet, or painter could pause from work to enjoy and admire unusual stones over a cup of tea or wine with friends. This appreciation of stones spread from Imperial China to other countries, likely during the Tang (618–907) and Song (960–1279) dynasties. Passionate connoisseurs such as Mi Fu (1051–1107) developed criteria for appreciating stone, laying the foundation for evaluating them. Mi Fu developed a rhyming list of the four characteristics of an excellent stone: *shou* (瘦, thinness), *lou* (漏, channels), *tou* (透, holes, or openness), and *zhou* (皱, wrinkles). The stones that displayed some or all of these qualities were more highly valued. Although newer criteria based upon shape, color, surface texture and patterns, material, and appropriateness of the base, have been adopted to accommodate the much wider range of stone types collected today in China, the original ancient criteria are still respected and applied, particularly to traditional stone types: Taihu, Lingbi, Kun, and Ying stones.

To facilitate indoor display of small- and medium-sized stones, support was needed to hold the stones and to prevent damage to tables. The hand carved wooden base evolved for this purpose. Regional styles of bases developed, sometime with recurring motifs. Stones were also mounted in incense burners and in bronze trays with smaller stones to hold them in place. In China, displaying stones in sand-filled ceramic trays is a modern technique adopted from Japanese traditions.

Asian Stone Appreciation in Transition

History abounds with examples of cultural adaptation: One group borrows an art, belief, or activity from another group and slowly, over time, shapes it to conform more comfortably to the ideals and practices of its own culture. Chinese stone appreciation initially had a significant influence on Japanese stone culture; that largely disappeared as the Japanese refined imported ideals and aesthetics, and moved from importing Chinese stones to searching out and harvesting local stones from the many rivers and streams of their native islands. Stone appreciation cultures in both Japan and Korea gradually diverged from their Chinese origins and became distinct branches of Asian stone culture. All influenced the fledgling stone culture in North America.

The transition in Japan from an interest in Chinese stones to a greater emphasis on local stones appears to have occurred in the Meiji Era (1868–1912). During this period stone exhibitions often contained a combination of Chinese and Japanese stones. The Chinese stones were referred to as *kai shi* (*guai shi* in Chinese, literally "strange stone") according to Hideo Marushima, the stone historian cited earlier. Stone historians in Japan have clearly documented that many of their most valuable stones originated in China. A 1903 Tokyo exhibit of bonsai and stones was the subject of the two-volume *Juraku-kai Zuroku*. This illustrated catalogue provided descriptions and detailed line drawings of the suiseki, bonsai, ikebana, and scrolls installed in the *tokonoma* (decorative alcoves) of two upscale restaurants. Three stones in this work were referred to as Chinese—two Lingbi stones and one undesignated Chinese stone. The 1909 work *Collection of Bonsai and Unique Stones* by Ebara Shunmu described the use of Chinese Lingbi stones in Japan and provided an illustration of one.

Marushima also described an exhibition of various Chinese and Japanese stones by a Kyoto club in that city in 1910, and noted that a copy of the famous Chinese work *Plain Garden Stone Catalogue* (*Suyuan Shipu*) was displayed with them. This work was first published in 1613, and reprinted many times. The Matsuzakaya Department Store in Ginza, Tokyo, worked in cooperation with the Arts and Crafts Public Company of China to display over a hundred Chinese stones, including Taihu and Lingbi stones, from February 7 to 12, 1967. Later, at least sixteen Chinese stones were included among Japan's historically important stones illustrated in the 1988 *Densho-seki* (*Historical Stones*) by Teisuke Takahashi. This is but a sampling of the documentation of the Chinese influence on Japanese stone appreciation, and indicates a long period of mixed traditions, during which Japanese stone appreciation gradually developed into a style of stone culture distinct from the Chinese.

Stone Appreciation in North America

Modern stone appreciation in North America (as distinct from that of Native Americans; see Paul Gilbert's essay) is directly linked to traditions of

China and Japan, less so to those of Korea. It is difficult to trace with any degree of accuracy when the first East Asian viewing stones arrived in North America; however, two separate main lines of transmission can be identified: the importation of antiques, including stones, from China, and also the activities of small clubs interested in various traditional Japanese arts.

Great quantities of Chinese art and antiques were brought to North America during the heyday of China trade in the Qing dynasty (1644–1911), and especially near the end of the dynasty when order in Imperial China was rapidly deteriorating and the nation was heading toward a lengthy civil war. Decades later, vast amounts of antiquities were taken from mainland China during the final years of the Chinese Civil War, effectively ending in 1949 with the establishment of the People's Republic of China. Most of those treasures were taken to Taiwan with adherents of the Nationalist government, but many also found their way to the West, including examples of old Lingbi, Ying, Taihu, Yellow Wax, and other stones. These stones were associated with traditional Chinese arts and crafts and thus were sought after by collectors and museums assembling collections of East Asian art. As a result, most Westerners learned about these stones from the catalogues of museum curators, complemented by those of auction houses and antique and stone dealers.

Over the past several decades major exhibitions at noted art museums and galleries have not only raised awareness of Chinese stones in North America and Europe, but have also clearly identified them as works of art. One of the earliest exhibits, *Kernels of Energy, Bones of Earth*, was held at The China House Gallery in New York City in 1985 and 1986. John Hay published a catalogue of the same title to accompany this exhibit. Following this, several major exhibits of Chinese stones were staged at important art museums in the United States, Switzerland, Germany, and France. The Richard Rosenblum collection was the first major group of Chinese stones to be featured in an exhibit mounted in the United States, and then traveled to Zurich and Berlin before returning to North America. The catalogue *Worlds Within Worlds* was prepared by Robert Mowry in 1997 to accompany this very important exhibit, which exposed many people to Chinese stones for the first time. Rosenblum's collection was ultimately shown at seven venues in the United States and Europe. In 1999, the prestigious Art Institute of Chicago staged the exhibit *Spirit Stones of China*, featuring the collection of Ian and Susan Wilson, and published a catalogue by the same name edited by Stephen Little, Pritzker Curator of Asian Art at the Institute. The status of Chinese stone appreciation in Western countries made another major advance when the Musée Guimet in Paris staged a major exhibit from March 28 to June 25, 2012.

Concurrently, several Asian antique dealers and purveyors of Chinese stones also made important contributions to the developing awareness of Chinese stone appreciation. Foremost among these is Kemin Hu, author of *The Spirit of Gongshi* (1998), followed by *Scholars' Rocks in Ancient China* (2002), *Modern Chinese Scholars' Rocks* (2009), and *The Romance of Scholars' Stones* (2011). Through her galleries, lectures, and online sales, Hu has been one of the primary sources of Chinese stones for American and European collectors. In 1995, Paul Moss, Brian Harkins, and other contributors published the catalogue *When Men and Mountains Meet: Chinese and Japanese Spirit Rocks*. More recently, Marcus Flack, a London-based Chinese antique dealer published the impressive *Contemplating Rocks* in 2012.

It is difficult to determine precisely when examples or information about Japanese indoor stone appreciation reached North America. In 1935 and 1936, Dr. Jiro Harada delivered a series of lectures on Japanese art and culture at several universities in the United States. One of those lectures, "Japanese Gardens," touched on stone appreciation. In his 1937 book *A Glimpse of Japanese Ideals*, Harada described three forms of stone appreciation. One was *ki-seki* (strange or unusual stones) placed on writing tables or in tokonoma "to lead the gazer into reveries from fancies suggested by their shapes, colors, and markings." He described *sui-seki* as "water stones" displayed in pottery trays or shallow basins with water; these stones were watered and great appreciation was shown when moss began to grow, creating landscapes with forests and meadows, sometimes with streaks suggesting streams and cascades. Harada also listed the use of rocks in trays with low growing or miniature bamboo grass to emphasize the immensity of nature.

This third form of stone appreciation, the tray landscape, appears to have been a more popular practice in Japan than suiseki prior to 1950. Soen Yanagisawa published *Tray Landscapes* (*Bonkei and Bonseki*) in 1955 as part of a series of English language books on Japanese culture; however, this series did not include a volume on suiseki. *Bonkei*, as defined by Yanagisawa, is the art of portraying a bit of scenery in three dimensions on a rectangular or oval tray made of metal, porcelain, or wood, while *bonseki* is the art of representing a landscape or seascape artistically on a black, rectangular, or oval lacquered tray or board with stones and sand. *We Japanese*, a popular book with numerous editions between 1934 and 1950 whose purpose was to educate Western visitors about Japanese customs, also devoted a full page to the art of bonkei while omitting any mention of suiseki.

Japanese stone appreciation reached North America primarily through hobbyists in some of the Japanese American communities in California, where local club members spoke only Japanese. Japanese American Richard Ota published two in-depth articles in the local Japanese-language newspaper *Rafu shinpo*, the first appearing on January 1, 1961, the second on January 1, 1968. These articles coincided with the peak of stone appreciation in Japan. Japanese Americans in the Oakland, California, area were also among the first to begin practicing suiseki in North America. One of the first major exhibits of Japanese stones was held at the Los Angeles County Natural History Museum in 1973, when nearly two hundred stones were displayed. This was part of a cooperative effort of the Los Angeles–Nagoya Sister City affiliation. Word of this interesting hobby began to spread and soon non-Japanese interested in Japanese arts and culture were attracted to collecting and appreciating natural stones. In 1983, the California Aiseki Kai club was established as a subdivision of the California Bonsai Society. This club, led by Larry and Nina Ragle, is devoted to the collection and study of North American stones following suiseki guidelines. California Aiseki Kai has been a source of information about Japanese-style stone appreciation through their monthly newsletter and annual exhibit held at the Huntington Library, Art Collections, and Botanical Gardens in San Marino, California.

On the Eastern Seaboard, Yuji Yoshimura, a Japanese bonsai master living and working in the New York area, was instrumental in promoting Japanese suiseki. Yoshimura took a group of East Coast residents on a trip to Japan in 1968 to help them learn about bonsai and suiseki. His meeting and discussion with Keiji Murata, publisher of *Juseki* magazine, is recorded in the October issue. Yoshimura later co-authored with Vincent T. Covello *The Japanese Art of Stone Appreciation*, published in 1984. Yoshimura, Covello, William N. Valavanis, and other East Coast colleagues organized the first national stone appreciation meeting, held in Rochester, New York, from June 10 to 11, 1994. This led to a series of international stone appreciation symposia held every other year in Hershey, Pennsylvania. These symposia have primarily emphasized Japanese-style stones and have been a major source of information for many. Another East Coast leader, Jim Hayes, formed The North American Viewing Stone Society and began publishing a quarterly magazine, *Waiting To Be Discovered*. Fifteen issues with articles on Chinese, Japanese, and American stones were published from 1996 through 1999, when it ceased publication.

As interest in stone appreciation grew in North America, more books appeared. An early book to feature North American stones, *Suiseki & Viewing Stones: An American Perspective*, was published by Melba Tucker in 1996. Most of the stones illustrated were collected by the author in California, largely from the desert regions of the south. Felix Rivera published *Suiseki: The Japanese Art of Miniature Landscape Stones* in 1997. Privately published books have also appeared, including Bill and Julie Hutchinson's *Suiseki in British Columbia*, published in 1976, and James L. Greaves' *American Viewing Stones, Beyond the Black Mountain: Color, Pattern, and Form* (2008). This book featured North American stones exclusively from his private collection.

In addition to published information, Nippon Suiseki Association Chairman Arishige Matsuura made occasional visits to North America to give illustrated lectures on Japanese suiseki principles. Matsuura helped to spread Japanese-style stone appreciation, especially among those practicing the art of bonsai. The fact that few, if any, Chinese stone specialists traveled to North America from China to teach and promote Chinese-style stone appreciation contributed to the preponderance of the Japanese style.

Thus twin parallel influences—Chinese and Japanese—of Asian stone appreciation in North America have helped foster an awareness of native stones as art objects. As this interest has grown, however, collectors have found that favored specimens do not always fit into Asian aesthetic standards; or they have difficulty deciding whether to make a Chinese or Japanese style base for their stones. It is understandable that they would seek ways to meld native stone display with native precepts and the American culture that has resulted from the blending the indigenous with the cultures from generations of immigrants. The time is ripe for a new paradigm to emerge.

Toward a New North American Paradigm

A modern North American framework for the appreciation of viewing stones should acknowledge East Asian precedents while at the same time encourage creative responses to the geological and cultural diversity of the New World. The stones presented in this book are owned both by adherents to traditional criteria and those favoring innovation; yet all are open to a more expansive view of what constitutes an excellent viewing stone presentation. In fact, the criteria used for selecting stones for this book may serve as guidelines for evaluating North American stones in general. The seven criteria used are listed below.

Overall impression is the impact that a stone and its base have on the viewer. This may provoke an immediate reaction, such as astonishment at a levitating form like the one at right above, or one that develops through studying the stone over time. What message does it convey? What image does it conjure? Some stones evoke such mixed reactions that they remain unnamed.

Creativity is demonstrated both in selecting a stone and in successfully pairing it with an appropriate base to create an object of aesthetic merit. For North American stones both activities continue to move well beyond the bounds of traditional East Asian styles and materials, as seen in the seamless presentation of *Building Supply*.

Unnamed, see Gallery 131.

Building Supply, see Gallery 55.

Water Poem, see Gallery 60.

Autumn on the Mountain, see Gallery 81.

Shape is critical, perhaps the most immediate contributor to overall impression. A stone may be reminiscent of a landscape, a real or mythical figure, or an abstract sculpture; it may exhibit obvious motion or a balance evoking serenity, as does *Water Poem*.

Color is also immediately apparent and so must support shape and texture to present a coherent and impressive whole. The color of *Autumn on the Mountain*, for example, decorates an ordinary mountain stone with brilliant fall foliage. Color may enhance a pictorial element, as a white quartz vein creating a mountain rivulet, or may simply be impressionistic, ranging from subtle and dark to cheerful and bright. North American stones have an infinite range of colors, from the serpentines and jades of the west to the beautiful agates gleaned from ancient gravel deposits in the Great Lakes region.

Texture includes both the rugged or smooth appearance of the stone surface, or discernible patterning, whether abstract or pictorial. Texture can lend an ancient or exotic character, as the time-ravaged cragginess of *Dragon Bone Stele*. It should complement other stone qualities, and no one texture is considered superior to another.

Material refers to composition, although collectors are generally more interested in stone appearance than with specific geological makeup or formation process. Petrified wood, such as *Tsunami*, and other fossilized organic forms, even meteorites, are of interest as long as they meet other selection criteria. In general, harder stones are favored simply because they are more durable.

Base refers to the platform used for display, with the materials and design often dictated by the stone, as seen in *Arch at Rattlesnake Canyon*. As the created object, the base should complement the found object, the stone, to present an artistic whole with impact on the viewer greater than the sum of its components. Creativity in designing bases promises to make a significant contribution to North American viewing stone aesthetics.

Collection and Presentation

Across North America today hundreds of collectors are searching for and finding excellent stones. Once moved from their natural locations, most need to be cleaned to some extent. Stones removed from fast flowing clear rivers and streams generally need less cleaning than those removed from muddy waters or dug from the earth. Most need organic material such as algae,

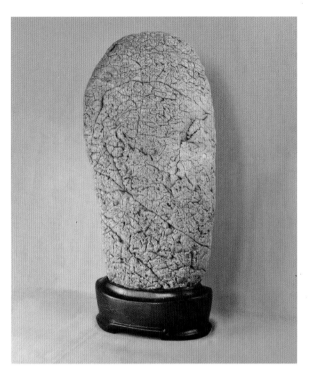

Dragon Bone Stele, see Gallery 43.

Tsunami, see Gallery 84.

Arch at Rattlesnake Canyon, see Gallery 64.

mud, and silt removed. This can be accomplished with water under pressure and soap, although sometimes vinegar, weak acetic acid, or muriatic acid is used. Stones removed from the earth may need additional cleaning with a wire brush. Although the surfaces of the stones should be cleaned, the natural shapes of the stones should not be modified.

A completely natural, found stone is considered best; however, strict adherence to this standard would be too limiting. A single baseline cut can often result in a more aesthetically pleasing stone than as originally collected. Mechanical cutting or altering the base of the stone to make it flatter is sometimes done in North America. This is also well-documented practice in Japan, particularly in the 1960s through the 1980s, and in China, especially for large, top-heavy, vertically oriented stones. In his *American Viewing Stones*, Jim Greaves states: "The prevalent American approach, while considering the unaltered, natural stone to be ideal, tends to accept the use of a single baseline cut." Many attractive landscape stones have resulted from a single baseline cut or altered base. Cutting stones to make a flatter base may lower the value of a stone in the eyes of some collectors even while increasing aesthetic appeal.

Stones may be polished to remove matrix material or expose mineral formations with interesting patterns. This was done with some Chinese and Japanese stones, especially so-called peony and chrysanthemum stones, still considered "natural" by collectors in both countries. Polishing leads to a wider suite of stones that can be appreciated; for example, a dull piece of American petrified wood that has been polished but retains its natural shape can make an admirable stone. While some collectors avoid this practice, others favor more flexibility in practice. North American stone appreciation should not be a rigid pursuit with inflexible rules, but rather an activity to be enjoyed within flexible guidelines.

Another controversial practice is the application of oils, waxes, or a combination of both to enhance the colors and to develop a patina; some believe this helps preserve the stone while others maintain it is destructive. Softer porous stones, including many calcium carbonate based stones, can be permanently altered with the application of oils and waxes. Harder silica-based stones will not absorb the wax or oils like the carbonate stones. Oddly, some purists who condemn this practice will sit for hours rubbing perspiration from their heads and hands onto a stone until the surface has been slightly altered, perhaps considering this more "natural." In any case, waxes or natural oils should always be used sparingly. The natural beauty of a stone is always best.

Orientation of a viewing stone is critical in presentation. Many stones have an obvious front or primary viewing position, while others can be appreciated from different angles. Some can even be rotated 360 degrees, with each degree of rotation changing the image presented. Thus it is best not to approach a stone with the idea that it must have a definite front and back. Some stones, of course, can only have one orientation, e.g., a stone that suggests a waterfall can only be positioned to show water falling to the lowest point. Other stones, particularly with abstract shapes, may have more than one impressive orientation. In such cases, multiple bases can be made, one for each orientation. For an example of this see "Positioning the Stone" in Kemin Hu's *Modern Chinese Scholars' Rocks*. The collector should always take time to study a newly acquired stone before rushing to have a base made. By placing a stone in one position, observing it for several days, then repeating this process, he or she can become confident of the orientation that presents the finest effect.

Hand-carved wooden bases, with styles definitely linked to specific East Asian countries or regions, remain the most common method of supporting and displaying stones. The bases of most stones presented here have been influenced by Japanese rather than Chinese styles. Several North American wood carvers have learned the Japanese way of making bases; unfortunately, few are capable of making quality bases in the more elaborate Chinese styles.

Shallow ceramic or metal trays, usually filled with sand, are often seen supporting stones resembling landscapes—especially mountain, island, and coastal scenes. Metal trays are usually made of bronze or copper, and are known most commonly by their Japanese name *doban*; shallow ceramic trays may be unglazed or glazed, and are frequently called by the Japanese term *suiban*.

As with other aspects of collection and presentation, Northern American collectors are becoming more receptive to more effective ways of supporting and displaying stones. Some approaches utilize fabric or small pillows or cushions. Other less traditional approaches are blocks, or slabs (e.g., driftwood), or thin, mat-like slices of wood. It is appropriate that contemporary North American stone appreciation framework should be open to a range of forms and base materials. However, note that none of the "new" attitudes or practices we have discussed are in opposition to traditional views that: 1) the more natural the stone the better, so less intervention is preferable; 2) creativity in stone appreciation derives as much from presentation as collection; and 3) any base should complement the stone it supports, and so together convey an impression more powerful than either. As perhaps best stated by Hao Sheng in *Fresh Ink: Ten Takes on Chinese Tradition*:

Any man-made interventions in the life of a rock—moving it to a new location, naming the rock, fitting it into a stand—should demonstrate creativity and contribute to a narrative about the interaction between man and the natural world.

THE RICH HERITAGE OF NORTH AMERICAN VIEWING STONES

Paul Gilbert

Viewing stone appreciation is a growing practice in North America, with participants generally acknowledging the influence of Japanese arts such as suiseki or Chinese fascination with scholars' rocks. While there is much to learn from these East Asian traditions, it would be a mistake to ignore the role that stones have played in the lives of our North American forerunners, the Native Americans.

Stones were very much a part of the fabric of the ancient civilizations of the Americas. They were collected not only for utilitarian purposes such as tools and tool making, but were also carefully chosen for their beauty, shape, and color. Other stones were credited with supernatural powers, valued for medicinal purposes, or revered for their resemblance to human or animal figures, sometimes worked into such images.

The ancient peoples of North America believed strongly in connections between specific landscape features and shared historical or religious narratives, passed from generation to generation through oral traditions. Ceremonial grounds were often marked with prominent stones, some standing, some perched on other rocks, and others balanced. The New England landscape is dotted with such sites, many unnoticed or thought to be the work of early farmers or settlers. Man-made piles of stones, cairns, were created as early as 5000 years ago in North America. Evidence suggests that along with stone walls these compact and carefully built cairns were used to demarcate ceremonial areas. Absent from these structures were left-over stones, leading archeologists to believe that each stone was a personal offering object.

Sacred architecture created by indigenous North Americans in the western United States includes medicine wheels, or sacred hoops. Stones were arranged on the ground in a circular pattern with lines forming spokes in all directions. The largest of these, in the Bighorn National Forest in Wyoming, contains 28 spokes, possibly indicating the days of a lunar month and used as a calendar by Plains Indians.

The winter solstice was perhaps the most important day of the year to Native Americans, particularly after agriculture developed, as crops were planted based on the position of the sun. Shadows cast by standing stones aligned with markers placed to indicate the winter solstice. Prominent, securely anchored standing stones appear to have been located carefully both to mark solar events and to signal topographical features, particularly the confluence of streams, or perhaps a spring that might be the origin of a seasonal tributary.

Ezra Stiles, president of Yale University from 1777 to 1795, an eminent American theologian, lawyer, and philosopher, was one of the early observers of Native American stone structures, and his writings are an original source of reference to Indian stonework. Stiles described about twenty worked stones that he called stone gods or "godstones," considering them to be idols. One of them he described as about three feet high and about one half to one foot thick, and carved to present a human appearance. Since Stiles' first observations, excavations of standing stones erected by Native Americans invariably reveal bedrock modified by humans, with grooves or niches carved to form a supporting base. Standing stones were also supported by being wedged into a socket in the bedrock. It is not known whether the bedrock was naturally exposed when it was worked, or if a great amount of excavation was accomplished to gain access to it.

Stiles observed and described stone mounds and large boulders on which Indians cast stones or pieces of wood as donations every time they would pass. In 1762 he writes, "The Indians being asked the reason of their custom and practice, say they know nothing about it, only that their fathers and their grandfathers, and their great-grandfathers did so, and charged all their children to do so; and that if they did not cast a stone or piece of wood on that stone as often as they passed by it, they would not prosper, and particularly should not be lucky in hunting deer."

Stones were used as personal offerings to the spirits. Stone piles and stone walls created by Native Americans often included niches or small open spaces to hold these. Additionally, archeologists have discovered perched boulder structures where a table rock or quarried flat slab is set on three or four supporting rock pedestals.

Typical Native American sacred stone structure, with openings for stone or grain offerings. Photo © Mercedes Soledad Manrique, from Dreamstime.com.

The open space underneath the slab was used as a niche for the placement of stone offerings. These might have been picked up in the vicinity of the wall or perched boulder, or, more likely, they would have carefully selected from a streambed or a mountainside and carried a considerable distance.

There is evidence that freestanding portable stone art objects were collected by Native American shamans. The shaman was the spiritual leader in early American societies and was believed to have access to and influence in the spiritual world. The shaman would use collected small figure stones during healing ceremonies. In the basins of western Nevada, 133 such freestanding sculptures of stone, bone, and wood have been discovered; most resemble various animals: rabbits, grasshoppers, owls, mountain sheep, and fish. Others appear to be abstract, utilitarian, or phallic in form. Natural or worked image stones were believed to bring luck and success. Early Puritan missionaries preached to Native Americans against idols and godstones. In the Great Awakening of 1741, portable stones and wooden art objects were brought and given up to the missionaries. These objects, considered evil, were no doubt destroyed.

Stones were also collected by the common people of ancient times. A stone of a special shape or beautiful color might be chosen to be placed at the grave of an ancestor or famous warrior as an act of reverence, remembrance, and honor. Piles of stones or wooden sticks were used to mark not only graves, but scenes of tragedy. A special stone was often attached to a hunter's bow as it was thought to help bring success on the hunt. Stones in the shape of an animal were carefully carried year after year, hunt after hunt, in a pouch or pocket to bring good luck.

Large stones and stone walls in canyons and flat surfaces on cliffs served as the canvas for artistic expression for early American civilization. Images were sometimes placed so that a small, apparently insignificant nodule or protuberance forms the eye of an animal, or other part such as the head or shoulder of a buffalo. Carvers used natural features of the stone surface to depict animals or human body parts.

Stones resembling the human figure were highly prized and collected. Those specifically shaped with a neck, shoulders, and torso, were called Manitou stones, a term that can be roughly translated as "spirit." These stones have been found at a number of archeological sites. Natural, un-worked, Manitou stones were especially prized; it appears that extensive working may have been sacrilegious.

Manitou stone, Newberry, MA. Photo by James Gage, from *A Handbook of Stone Structures in Northeastern United States*.

The size of Manitou stones varies but not their general shape. Many were small and portable; others ranged up to three feet and were permanently placed, sometimes in rows. These anthropomorphic stones are found throughout North America as a component of Indian ritual. There is evidence that the portable stones may have been taken into individual dwellings, with the larger stones placed on hills or other high points associated with water. Although springs seem to have been most important, Manitou stones have been used to mark waterfalls, rapids, headwaters, river bends, creeks, marshes and drainage divides.

Quartz was one of the most esteemed and collected stone types by Native Americans and quartz pieces were referred to as lightning stones. Puebloan farmers in the Southwest called them thunderstones, believing them to create rain. This name comes from the fact that when two pieces of quartz are briskly rubbed together in the dark, a physical property known as triboluminescence can produce a flash of light inside the stone resembling lightning in the clouds. This was seen as a supernatural manifestation and was incorporated into religious worship common throughout the Americas. To best witness the flash, the quartz would be taken into a dark, round subterranean room for ceremonial worship called a *kiva*.

Another finding to emerge from North American stone research is that petroglyphs were often pecked or hammed using unmodified quartz cobbles as hammers. A high concentration of broken quartz has been found at rock art sites. No doubt the triboluminescent property of the quartz caused the hammer-stones to glow, especially at night, and added to the belief of the supernatural power of collected quartz. It also appears that leaving behind a valued stone, especially quartz, as an offering to the gods at petroglyph sites was very common.

Chinese stone collectors from ancient times shared with Native Americans a love of turquoise. Native Americans were working turquoise mines with stone mauls and antler picks for centuries before the arrival of Europeans. To the native people of the American Southwest in particular, turquoise, known as "sky stone," was an ancient talisman for health and happiness and for centuries was incorporated into ceremonial rituals, jewelry, sculpture, and pottery. Although turquoise was not mounted into silver jewelry until 1890, before it was collected, carried, displayed, traded, worn, and highly valued. The Navajo even exchanged turquoise as their currency. The Zuni people of western New Mexico valued it most of all—a string of turquoise beads could be worth several horses.

While turquoise artifacts are extremely rare, examples of worked turquoise dating from the first to third millennium BCE were found during initial archeological excavations of the Hohokam site along the Gila River in Arizona. Ancestors of today's Zuni people carved turquoise into good luck charms in the shape of wolves, coyotes, or mountain lions, as they were thought to bring luck. Turquoise was an important mineral throughout prehistoric times, as evidenced by the 56,000 pieces, mostly in the form of beads and pendants, found in a single burial site during an 1890s archeological dig at Pueblo Bonito, in Chaco Canyon, New Mexico. Native Americans believed that turquoise would let the dead rest in peace and it was carved, usually into animal shapes, which were then placed into graves.

It is impossible to study the early civilizations of North America without understanding the importance and value of stones. They were held in deep appreciation not only for their practicality, their utilitarian value in hunting and gathering, and for the religious beliefs attached to them, but also for their beauty, color, and form. Although there is a wealth of information about the history of viewing stones in Japan and China, there is much less available regarding stone collecting in early North America. Nevertheless, today's North American Viewing Stone collector should never forget that on this very continent we were preceded by countless generations of stone lovers, the Native Americans.

NORTH AMERICAN STONE APPRECIATION CONCEPTS

Rick Stiles

The history of mankind could be written as a history of man and stone. We evolved from a Stone Age people; stones were our weapons, tools, monuments, and amulets. Archaeological evidence for this exists throughout the world, and no single nation or culture holds exclusive claim to this heritage. Our contemporary pace of life, with all of its digital distractions, may obscure these ancient origins but it cannot impair the deep connection with stone universal in the human experience.

North American stone appreciation draws on the heritage of distinct Chinese, Korean, and Japanese stone traditions. These traditions developed under the influence of history, philosophy, spirituality, culture, geology, and landscapes in their countries of origin. Each of these Asian traditions features distinct characteristics that are recognized by connoisseurs today.

In contrast, contemporary North American experience with stone appreciation is less than 100 years old. Asian immigrants brought the activity to North America early in the twentieth century. They organized societies for stone appreciation. To some degree, these pioneers stood outside the mainstream in their new land so this interest provided them with a cultural anchor. They strove to be faithful to their heritage. The whole point of their activity was to nurture traditional forms, or more precisely, what they believed to be traditional forms. They would have seen no reason to develop a new and distinct North American stone appreciation tradition.

However, once migration occurs, change can be expected. Conditions in the new location alter what is imported; dogma weakens over time. The shift may be imperceptible at first, because cultural drift can be subtle. As time passes, principles become unlinked and, eventually, the import assumes a local flavor, a native character. Within a generation or two, it begins to take shape as a new tradition. Now that stone appreciation has migrated to North America, one can be confident that a distinct North American stone appreciation tradition will emerge. No one will ring a bell when this happens, but it will be obvious in hindsight. Future scholars will be able to articulate a clear understanding of its origins, and the images in this book will be important source material for their work.

Seiji Morimae, a Japanese teacher, once offered the gentle comment that he would like to see students learn the basics of Japanese suiseki and then adapt stone appreciation to their own circumstances. He said, "When a Japanese person looks at images of the American West, they don't completely understand." He offered the thought that a while a Japanese master might understand a hut stone, he might not grasp the full meaning of Laura Ingalls Wilder's *Little House on the Prairie*. The rich cultural context of Wilder's simple children's stories—the fate of Native Americans, the influx of pioneer immigrants, the doctrine of Manifest Destiny, and the closing of the frontier—might make little impression on a non-American reader. Why North Americans were not more focused on being North American seemed to puzzle Mr. Morimae.

Future scholars will likely say that the development of a distinct North American stone appreciation tradition was driven by three powerful enabling factors. The first of these is geology. Rocks and minerals are not evenly distributed throughout the world. Formation, erosion, and weathering processes vary from place to place. A large country like China produces many stone types, while smaller countries offer less variety. North America has a striking advantage because it is a huge continent with diverse geology; its stones have been formed by a wide variety of processes and shaped in a wide range of conditions. Those in this book demonstrate astonishing variety and quality. Many are unique.

Geisha Girl, aka Geisha, see Gallery 49.

The stone above, collected by Melba Tucker, was first described in her book *Suiseki and Viewing Stones: An American Perspective*. This type of multi-colored jasper is popularly known as Indian blanket stone, and is a prime example of the stunning variety of North American material. In her book, Tucker named this stone *Geisha Girl*, an odd choice revealing her orientation toward what

now must be considered an unauthentic version of the suiseki tradition. At the time little thought was given to the idea that a distinct North American tradition could have merit. This stone is obviously North American. The geology of Japan could never produce a stone like this, and there has never been a suiseki that looked anything like it. It was saddled with a misnomer because of a bias that a North American stone could have value only if perceived to have some Asian connection.

With the passage of two decades, this remains a great stone. It is a brilliant and rare North American stone with the embedded image of a small tree struggling in our western mountains. Of course, the tree may also be seen as the *obi* on a *kimono* but that is a forced observation. Posterity will re-position this stone. History will be re-written. This stone will be seen as a poignant marker of the moment in time just before North American stone appreciation changed.

The second enabling factor for a distinct North American stone appreciation tradition is landscape. Chinese literati idealized the notion of the scholar-hermit, living a contemplative life in remote mountains. This ideal was intimately connected with landscapes that actually existed, such as the limestone karst terrain of the Li River. When natural stones from such places were brought into the private space they became representative of the conceptual ideal. The actual landscape was not incidental; it was more than just scenery. It affected the thinking, and reinforced the notion that a preferred life should be lived in concert with nature, removed from the clatter and corruption of "civilization." If such mountain landscapes had not existed, or had they not been idealized in this particular way, the distinct Chinese stone appreciation tradition would not be what it is.

North Americans have preserved and protected some of the most spectacular landscapes in the world under public stewardship. National parks and monuments are everywhere. However, our concept of landscape is something more than visual experience. Landscape is the stage on which history happens. Landscape also confirms individual identity and connects individual purpose to a broad understanding of the world. In this respect, Native Americans were conceptual thinkers just like the Chinese literati elite. Landscape was sacred to them. It carried profound spiritual and philosophical meaning.

The early nineteenth-century Hudson River School of American painters placed man in an idealized relationship with nature, although by then the frontier was already distant from them. By the close of the same century, wilderness was seen as something to be subdued through exploitation and settlement. Native populations were to be swept aside. Forests needed to be cut, and gold needed to be mined. In time, this understanding changed, through an increasing awareness of what was being lost. Early in the twentieth century, Canadian landscape painters known as the Group of Seven initiated the first major Canadian art movement based on direct contact with the natural landscape. They illustrated the change in attitude by depicting the Canadian landscape as an unspoiled place.

Today, there is strong support for environmental preservation throughout North America. For many, the relationship between man and nature is held as sacred once again. Native American traditions now command great respect. The non-sectarian spirituality of environmentalism is embedded in mainstream culture, including the notion that the natural landscape is a precious resource whose fate is inextricably entwined with that of humanity.

Landscape has always been more than scenery in the American West and certain landscapes quintessentially Western American. At Canyonlands National Park and Arches National Park in Utah, sandstone rock has been cracked into parallel lines by the pressure of salt domes. These places have just enough rainfall to erode the stone and create delicate, improbably attenuated spans like the one recalled in the arch stone below from the Mojave Desert. The Anasazi, ancestral pueblo people, inhabited Utah lands just south of these national parks beginning around 700 CE. Their petroglyphs are everywhere. The story of the rise and fall of their civilization is not completely understood, but it may be cautionary. Climate change may have played a role and that echoes current environmental concerns.

The Chinese ideal of the scholar-poet meditating in a mountain pavilion fits poorly in

Living Arch, see Gallery 36.

this stern land. The pueblo was not a pavilion for elites to drink wine and write poetry. You pulled your ladders up at night. Doorways were T-shaped so that only one set of attacking legs could enter. Life in the American West was driven by survival, and conflicts over resources, especially water, continue to the present day.

Our North American landscape is different and our conceptual idea of landscape is different. Competition has often been harsh, and like this stone, a little rough around the edges. We hold the land dear. We carry guns. Landscape has shaped the character of North Americans, and that legacy is evoked when we gaze at this stone. This is not the Li River valley. This is not Mt. Fuji. This is more than casual scenery. This stone speaks volumes about the North American experience and presents layers of meaning for the contemporary North American viewer. This is the conceptual landscape of Georgia O'Keeffe, where cattle skulls float in penetrating ultraviolet light.

The final enabling factor for a distinct North American stone appreciation tradition is idiom. Many North American stones are capable of speaking to Chinese, Korean, and Japanese traditions with great clarity, and there are fine examples in this book. But a distinct North American stone appreciation tradition needs to discover and embrace a distinctly North American idiom. This begins at the critical point when a stone is first recognized and seen to be something more than the literal object that it is. It is interpreted as an object that speaks to the culture of the collector, and is placed in an artistic context within his or her distinct stone appreciation tradition.

In language, idiom is an expression that has a meaning different from the literal. The true meaning is culture-centric. One can understand the idiom if one has a native understanding, but for an outsider the task can be difficult. The power of idiom operates with forms of expression other than language. Many Asian artists have painted Song dynasty stone lover Mi Fu bowing to a stone as his "elder brother." On one level, this legendary scene may only seem to be a fanciful theme. However, the image is really a Chinese cultural idiom. The elder brother reference is neither casual nor comical. Elder brother-younger brother was a critical category of Confucian order and one of the pillars of Chinese society. The idiomatic meaning is different from the literal image. It speaks both to an acceptance of Confucian order and also to a vision of the natural world that affirms the validity of Confucian order. This would be obvious to anyone living in a Confucian society.

Stone appreciation can also be a vehicle for idiom. On one level, a Chinese scholar's stone can be admired as a miniature landscape of mountain peaks, caves, or waterfalls. It conveys meaning that is representational, literal, and accessible to all. A Chinese scholar might understand deeper idiomatic levels of meaning. The stone may evoke a journey to visit a recluse poet or perhaps to the mythical islands of Peng Lai, with caves inhabited by Daoist immortals. A stone is not simply a stone. Stones communicate complex meanings that are culture-centric.

The American bison is a classic North American image. Animal shaped stones can be thin of meaning, with images too literal. This stone is an exception. Following the last Ice Age American bison thrived in large nomadic grazing herds as a key element in the Great Plains ecosystem. Humans also crossed the Bering

Strait at some point, and the history of bison and humans became intertwined. However, one should note that modern horses did not arrive on the Great Plains until sixteenth-century Spanish explorers brought them in from Europe. Bison hunting from horseback could not have been part of Native American life prior to that time.

Familiarity with the idiom provides access to a widely shared set of cultural understandings that are not literal. For most North Americans, the idiomatic meaning of bison includes a shared mythology about the natural state of the continent prior to European settlement, and the loss of Eden thereafter. We fought over, sectioned off, plowed and settled the prairie. We nearly exterminated the American bison. The life of Native American tribes changed forever. Even if Eden may have been an illusion, this idiom has power. This bison crouched in his wallow, head slumped, evokes a full book of ideas.

Such a book would have many chapters. *Little House on the Prairie* is stuffed with cultural baggage. You have Buffalo Bill's Wild West Show, buffalo soldiers, buffalo nickels, Howdy Doody's

Sitting Bison, see Gallery 54.

sidekick Buffalo Bob, Frank and Deborah Popper's Buffalo Commons, and Buffalo Bills football with a big fat bucket of Buffalo wings. The American bison idiom is broad-ranging. The true miracle is that nearly all of this can be conveyed in an instant to the North American audience. They already know it, and when they look at this stone, they apprehend it all immediately.

If North Americans look only for stones that speak to Asian collecting traditions, they will miss a grand opportunity. North America offers incredible diversity. Our culture is a blend of so many influences. It has been nurtured by Native American traditions and wave after wave of immigration by people from everywhere. The sheer variety of our philosophical and spiritual interests cannot be matched, and our society, perhaps more than any other in history, has promoted individualism and innovation. This mélange guarantees that our stone appreciation tradition will be both expansive and unique. We should embrace this opportunity.

This past year, the Pacific Northwest stone appreciation community undertook serious discussions about mission. At the core of these discussions was the question of what it means to be a North American. Our members seem to prize our cultural traits of independence and autonomy. We seem to be striving for a distinct vision and style that will gather up geology, landscape, idiom, and artistic intent in one package. This has not evolved into a formal school or practice as yet, but it will happen.

Stones like Edd Kuehn's *Doppler Dazzler* illustrate this direction. This was the poster stone for our 2013 STONE IMAGES IV group show at the Pacific Rim Bonsai Collection pavilion in Federal Way, Washington. One might regard

Doppler Dazzler as the stone equivalent of Marcel Duchamp's 1912 painting *Nude Descending a Staircase*, which stunned the world when it was shown at the transformative 1913 New York Armory Show. One hundred years later *Doppler Dazzler* is a landmark stone that breaks every traditional Asian rule brilliantly. It might well be iconic for this moment.

Kuehn's stone is not limited by gravity. The sleek hardwood component levitates it into space where it floats weightlessly, reminding one of William Wordsworth's line "I wandered lonely as a cloud." This obvious allusion to one of the most famous poems in the English language is breathtaking. Any Song dynasty sage would be startled by it.

The name, however, was initially jarring to my ear. Maybe that should have been a clue. Landmark works are often disturbing. I couldn't understand how the Doppler reference made sense until I talked with Edd Kuehn about his artistic intent. His title actually anchors this piece right in the middle of contemporary North American life. Admittedly, we're a little obsessed with the weather here in the Pacific Northwest. In the past, people looked out the window to see what was happening. Not now. We go to our phones for hour-by-hour Doppler radar images. This work grasps exactly what it means to be a contemporary North American beset by digital distractions. In a brilliant way, it calls attention to the character of our current society.

The "dazzle" of this antique form speaks to the insularity of our own era. We are spellbound by our phones. Hardly anyone would ever pause to look at a stone. *Doppler Dazzler* snaps a finger at this contemporary viewer. Think about the long history of stone appreciation. Think about the progress of fine art through time. Think about the life you are living. Be dazzled by this North American object. It is beautiful and profound.

When future collectors look back at *Doppler Dazzler* and many of the stones in this book, I believe they will speak of the period when North American stone appreciation declared its independence.

Doppler Dazzler, see Gallery 101.

VIEWING STONES AND CONTEMPORARY ART

Richard Turner

We express our creative impulses in diverse forms and media—dance, drama, film, poetry, prose, painting, sculpture, and a host of other modes. The collection and display of viewing stones is yet another. If we think of the expressions of our creative impulses as a spectrum, we might then put the experience of the stone collector plucking a prize stone out of a mountain stream at one end of the spectrum and that of the artist in his studio fabricating a sculpture inspired by a Chinese scholar's rock at the other end. Underlying the experience of the collector on the trail and the artist in his studio are fundamental affinities that connect these two apparently distinct creative acts.

Let's begin by thinking about the artist not as a maker-of-objects, but as a guide, perhaps like a viewing stone enthusiast who is taking us on a hike to one of her favorite sites. She might tell us what kinds of stones we could expect to find and show us the best places to start our search. The artist in the role of the guide points us toward objects and experiences that we might not, at first, think of as art. He asks us to consider a different way of regarding the world around us. Pablo Picasso and Marcel Duchamp are such artists. Picasso's 1912 use of a print of chair caning in one of his paintings inaugurated the concept of the found object. Duchamp's ready-mades employed unaltered common objects such as a urinal that he designated as works of art. Picasso's collage and Duchamp's urinal are, to be sure, objects, but their significance does not lie in their object-ness as much it does in the ideas they embody and the directions that they lead us. Over the course of the twentieth century the use of found objects has become standard practice in the art world.

Stones are, literally, found objects. Submerged in a stream or protruding from the desert sands, they are no more works of art than the automobile tire and stuffed goat were before artist Robert Rauschenberg used them in his 1959 sculpture *Monogram*. We are asked to consider the stone, the tire and the goat art by means of the agency of the person who selects and then displays them. The art lies in the act of re-contextualization. Mounting a stone on a base is akin to bringing it into a museum or incorporating it into a sculpture.

Viewing stones and sculpture have more affinities than first meet the eye. In both East and West there is a spectrum of uses for stone that runs from stone-as-material to stone-as-object. At one end of the Western spectrum might be Michelangelo's carved marble *Pieta* and at the other Michael Heizer's granite boulder *Levitated Mass*. The Eastern spectrum might be defined by an intricately carved jade sculpture at one end and a scholar's rock dredged from the bottom of a lake at the other. There is, however, a component of the Eastern spectrum that has no counterpart in the West. The interventions of the viewing stone "sculptor" do not correspond to any Western practice. The Chinese or Japanese connoisseur or artisan is guided by a desire to work the stone as if the very hand of mother nature herself had polished the surfaces with rushing water, etched the forms with windblown grains of sand or shaped the stone's contours with eons of crushing pressure. Any traces of human intervention are discouraged. The Western artist may be inspired by windswept and water worn forms but he makes no pretense of acting in the stead of nature.

The artist working in his or her studio and the viewing stone enthusiast splashing up a mountain stream might not, at first, seem to have much in common. A closer look, however, reveals intriguing analogies. When in the course of a hunting expedition we find a stone that we might want to add to our collection, we instantly evaluate it. Do we put it in our pack or leave it behind? Whether or not it fits within an established category—a waterfall stone, a hut stone, a distant mountain stone—the criteria that we use to assess our find—form, texture, color, unity, variety, harmony, proportion—are the very elements and principles of design that have been used by Western artists for centuries. They constitute the essential vocabulary for discussion of the formal qualities of artworks.

The artist contemplating the next stage of a painting-in-progress and the viewing stone collector fitting a stone to a base also share related experiences. The decisions the viewing stone collector makes in the course of designing and crafting a base for a stone are, like those of the artist, primarily formal in nature. Where is the top of the stone, the bottom, the front and the back? What combination of angles—side to side, backward and forward—best displays the perceived qualities of the stone? What is the appropriate proportion of the size of base to the stone? Where should the feet of the base be located so as to best support the stone? Is the base to be natural or painted wood? What color is the stain or paint? For the collector, as for the artist, these decisions are guided by a combination of formal or informal training in the received tradition and individual intuition that matures with experience.

The North American viewing stone collector's practice is typically based on precedents codified in Japan, and to a lesser degree in China and Korea. Contemporary artists, like those before them, work with and against tradition. A work of art, no matter how defiantly original it may be, does not come out of nowhere. Study of the history of a practice, whether it is abstract painting or suiseki, informs the creative choices of both the artist and the viewing stone collector. Visitors to the Metropolitan Museum of Art frequently see artists with their easels set up in front of paintings such as Frans Hals' seventeenth-century work *The Merrymakers of Shrovetide*, working to capture the bravura impasto of the Dutch artist's brushwork. In another gallery of the same museum visitors can see the radical innovations of Jackson Pollack's 1950 drip painting *Autumn Rhythm (Number 30)* which is as potent a challenge to the existing order today as it was when it was painted. Traditions, whether they are oil painting or stone appreciation, need to be preserved and challenged, practiced faithfully, and constantly questioned; otherwise they become moribund and die out.

A clue to nurturing the nascent North American viewing stone tradition may be found in the hybrid model for the artist currently being taught in art schools. In addition to learning the requisite skills to begin a career as a producing artist, today's art student is encouraged to be a writer/critic and curator as well. The days of the brilliant but inarticulate artist who leaves interpretation of his work to the critic and the exhibition of his work to dealers are for the most part long gone. Most contemporary artists study critical theory, write statements about their own work and that of others, and then often band together with fellow graduates to show their work in galleries they run and for which they curate exhibitions. A similar amalgam of competencies might serve the viewing stone collector as well. Although exhibitions of viewing stones are mounted annually across the country, curatorial standards are inconsistent and innovation is rare. There is also a dearth of scholarly and creative writing about viewing stones. Collecting, for many, remains a hobby rather than a serious pursuit. There are, nonetheless, signs of change. A few committed collectors are working to establish permanent homes for their stones and those of others at recognized institutions. Adventurous individuals around the country are crafting innovative bases for stones that are, themselves, outside the established canon. A scattering of writers is examining the current state of stone appreciation with a view toward defining the rough outlines of current practice. Occasional exhibitions presenting viewing stones alongside contemporary artworks have been mounted in California and elsewhere. In the words of singer Bob Dylan "the times they are a-changing," and faced with the inevitability of change, a willingness to experiment, to interrogate established precedent, and to risk failure is where the world of the viewing stone collector and the contemporary artist meet.

A recent example of innovative curating is the 2011 *Structure and Absence* exhibition at the White Cube Bermondsey in London. Curator Craig Burnett displayed Chinese scholars' rocks together with contemporary works of art "to impart an atmosphere of freedom to the works that surround them, diverting our attention from meaning to material, tweaking our customary ways of looking at contemporary art." This kind of free-ranging interpretation, unhindered by thoughts of the artist's intent, the meaning of the artwork and its place in the history of art,

can be a boon to a novice museum goer. This is not to say that familiarity with the historic context of the artwork and the intent of the artist does not enrich our understanding of a painting or sculpture. Burnett's suggestion is simply an entrée to the sometimes-intimidating world of contemporary art. Such an approach is especially appropriate when it comes to abstract art. Although the terminology differs in East and West, the formal properties of stones and the formal qualities of abstract sculpture are related. Scrutinizing a stone can help us understand an abstract sculpture and vice versa.

Robert Mowry in his essay for *Worlds Within Worlds* makes this very point when he writes: "The formal similarities between scholars' rocks and modern abstract art are striking. The rocks' affinity to modern abstraction is not the invention of contemporary sensibilities. On the contrary, their inherent abstraction was part of their significance to Chinese collectors." It is precisely for this reason that contemporary artists seem to gravitate more towards Chinese scholars' rocks than Japanese suiseki. Another reason may be the fact that the aesthetic criteria for scholars' rocks, in addition to setting guidelines for evaluation of form, texture, and color, encourage appreciation of the grotesque, awkward, and ugly. These qualities, first proposed in the eighteenth century, have an uncanny resonance for many twenty-first century artists whose work deliberately tests the limits of conventional standards of "beauty."

The contemporary artist's inclination toward Chinese scholars' rocks rather than Japanese viewing stones is somewhat ironic given the fact that Japan has been the primary focus of the West's interest in the arts of Asia since the mid-nineteenth century and the Japonisme that intrigued the Impressionists. From the architecture of Frank Lloyd Wright to the midcentury modern furniture of George Nelson, the austere harmony of Japanese design has been an enduring influence on Western painting and sculpture. John McLaughlin's sojourn in Japan was the inspiration for his later minimalist paintings. Light and Space artists James Turrell and Robert Irwin both acknowledge the role of Zen Buddhism and the reductive Japanese aesthetic in their writings.

This is not to say that Western artists, writers and musicians have been exclusively interested in the art and culture of Japan. As is often the case when ideas travel through time and across space, concepts and cultures comingle. Jack Kerouac's *Dharma Bums* and Gary Snyder's *Cold Mountain Poems* are animated by the authors' enthusiastic, if undiscriminating, embrace of Chinese and Japanese literature and religion. The paintings of Northwest Coast artist Morris Graves unabashedly reference Chinese still life painting while those of his contemporary, Mark Tobey, celebrate his encounter with Japanese calligraphy. Like Tobey, Abstract Expressionists Franz Kline and Robert Motherwell were both admirers of the spontaneous brushwork evident in some forms of Chinese and Japanese calligraphy. The musician John Cage was inspired by the Chinese *I Ching* to introduce indeterminacy into his music and the choreography of his partner Merce Cunningham. He was also a student of Zen Buddhism. Many of these artists were featured in the 2009 exhibition *The Third Mind: American Artists Contemplate Asia 1860–1989* at the Guggenheim Museum. For twenty-first century artists creating work referencing viewing stones, the arts of China and Japan continue to be a wellspring of inspiration.

Unsurprisingly, several contemporary Chinese artists have rediscovered scholars' rocks. Zhan Wang, whose hollow, stainless steel scholars' rocks and garden stones can be found in collections around the globe, is one such artist. His decades-long explorations of the scholars' rock tradition has resulted in works that range from a floating *Island of the Immortals* on a barge temporarily moored off the coast of Denmark to a proposal to send a stainless steel stone into space to replace a fallen meteorite. Liu Dan does exquisitely detailed large scale drawings of stones using traditional brush and ink. Sui Jianguo employed computer technology to design a wireframe version of a rock that he then translated into a huge painted metal honeycomb titled *Dream Stone*.

Pioneering work by ceramic artists in China and the United States inspired by Chinese stones is regularly on exhibit in contemporary galleries. Why ceramics? Perhaps it is the fundamental material affinity of clay and stone, or it may be the fact that significant aesthetic and technical aspects of Western ceramic practice find their origins in China and Japan. Zhao Meng and Ming Bai create subtlety glazed porcelain sculptures that take the form of scholars' rocks. Keiko Fukazawa slip casts and fires stuffed Beanie Babies, stacking them in piles that evoke scholars' rocks that are almost phosphorescent in the intensity of their brightly colored glazes. Julia Kunin, Adrian Saxe, and Shoshi Kanokohata are also well known for their ceramic sculptures that summon thoughts of viewing stones.

Contemporary artists have come to viewing stones from a variety of distinct perspectives. Pop artist Roy Lichtenstein, in the latter years of his career, designed comic book versions of scholars' rocks to accompany his Pop Art renditions of Chinese landscape paintings. Brice Marden's painting series *Cold Mountain* was inspired by seeing a famous stone in a garden while on a trip to China, as well as by stones in his own collection of scholars' rocks. Swiss born artist Ugo Rondinone's towering sculptures directly reference Taihu stones. Wandering through a group of his oversize sculptures in a gallery, brings to mind thoughts of a Chinese garden or the Stone Forest in southern China. The well-known story of the poet Mi Fu reverently bowing before his *Stone Brother* directly inspired an Arlene Shechet sculpture in which she has paired a commercial figurine of a Chinese scholar with an amorphous lump of fired clay. Andrea Cohen has cast plaster against bubble wrap to produce flamboyant pastel versions of Chinese stones. Australia-based artist Benedict Ernst carves suiseki-like sculptures from recycled Styrofoam, mounting the pieces on traditional wood bases to enhance the artifice of the "stones."

These artists and many like them have "discovered" viewing stones over the last two decades. Their experiments with form, scale, and material have established a new perspective from which we can take a fresh look at traditional Chinese and Japanese practices. Of equal, if not greater importance, is the potential for the aesthetics of viewing stones to impact the larger art world. We have seen how an interest in Chinese and Japanese stones has influenced the work of a few ceramicists, sculptors and painters. Consider how a broad familiarity with the standards for stone appreciation, Chinese and/or Japanese, would provide artists and others with the opportunity for a new understanding of stones as objects and stone as a material. What

might be said of Michael Heizer's *Levitated Mass*, the 340-ton boulder installed over a concrete slot at the Los Angeles County Museum of Art in 2012 when it is scrutinized from this perspective? Would Tony Cragg's fiberglass sculptures or Joel Morrison's stainless steel pieces, both of which readily suggest rocks but were not inspired by the artists' interest in viewing stones, be differently understood or have a deeper resonance, if they were examined in light of the aesthetics that guide viewing stone appreciation?

A reverse example of the application of an Eastern aesthetic to Western artworks can be seen in the case of British sculptor Henry Moore (1898–1986) and the impact of his work on recent stone collecting trends in China. The flowing organic forms and prominent negative spaces of Moore's sculptures resemble rocks shaped by water and wind. So obvious is the correspondence between his sculptures and viewing stones that the Chinese have given the name Moore Stones to rocks that, like Moore's sculptures, have smooth surfaces, uniform colors and abstract shapes.

If a deceased Western modernist sculptor can influence the taste of a twenty-first century Chinese stone collector then surely the experimentation of contemporary artists can become an important resource for today's viewing stone collector. As the practice of viewing stone appreciation in the United States matures, the affinities shared by the stone collector and the working artist can serve to inform the definition of a truly contemporary viewing stone idiom.

REFERENCES AND FURTHER READING

Burnett, C., D. Gamboni, and J. Saltz. *Structure & Absence*. Exhibition catalogue. London: White Cube Publications, 2012.

Covello, V.T., and Y. Yoshimua. *The Japanese Art of Stone Appreciation: Suiseki and Its Use with Bonsai*. Rutland, Vermont & Tokyo: Tuttle Publishing, 1984.

Davidson, Abraham A. *The Story of American Painting*. New York: Galahad Books, 1974.

Dupras, D.L. *California Jade: The Geologic Story of Nature's Masterpiece*. Privately published, 2011.

Elias, T.S. and H. Nakaoji. *Chrysanthemum Stones: The Story of Stone Flowers*. Warren, CT: Floating World Editions, 2010.

Gage, M.E., and J.E. Gage. *A Handbook of Stone Structures in Northeastern United States*. Amesbury, MA: Powwow Books, 2011.

Gerstle, M. *Beyond Suiseki, Ancient Asian Viewing Stones of the 21st Century*. Arcata, California: Water Stone Press, 2006.

Gomez, D. "The Allure of Turquoise." Santa Fe, NM: *New Mexico Magazine*, 2005.

Greaves, J. *American Viewing Stones, Beyond the Black Mountains: Color, Pattern and Form*. Santa Monica, CA: American Viewing Stone Resource Center, 2008.

Guth, Christine. *The Art of Edo Japan: The Artist and the City 1615–1868*. New Haven and London: Yale University Press, 1996.

Hao, Sheng. *Fresh Ink: Ten Takes on Chinese Tradition*. Boston: Museum of Fine Arts, 2010.

Harada, J. *A Glimpse of Japanese Ideals: Lectures on Japanese Art and Culture*. Tokyo: Kokusai Bunka Shinkokai, 1937

Harrist, R.E. Jr., and W.C. Fong. *The Embodied Image: Chinese Calligraphy from the John B. Elliot Collection*. Princeton, NJ: The Art Museum, Princeton University, 1999.

Hay, J. *Kernels of Energy, Bones of Earth: The Rock in Chinese Art*. New York: China Institute of America, 1985.

Hayes, J. ed., *Waiting To Be Discovered*. Quarterly magazine published 1966 through 1999 (15 issues).

Hu, K. *Scholars' Rocks in Ancient China*. Trumbull, CT: Weatherhill, Inc., 2002.

Hu, K. *Modern Chinese Scholars' Rocks*. Warren, CT: Floating World Editions, 2006.

Hu, K. *The Romance of Scholars' Stones*. Warren, CT: Floating World Editions, 2011.

Hutchinson, B. and J. *Suiseki in British Columbia*. Victoria, BC: Privately published, 1976.

Jia, X. H., ed. *Phenology (History) of Chinese Stone Appreciation*, 2 vols. Text in Chinese. Shanghai: Shanghai Scientific & Technical Publishers, 2010.

Jeong, Cheol-hwan, et al. *The Bucheon Museum of Suseok*. Seoul: Bucheon Cultural Foundation, 2008.

Kim, J. *Viewing Stones, Korean Classical Concepts: The Juneau Kim Collection Reflecting Korean Literati Views*. Irvine, CA: Privately published, 2009.

Lowry, J.D., and J.P. Lowry. *Turquoise Unearthed*. Tucson, AZ: Rio Nuevo Publishers, 2002.

Martin, P.S., in Sturtevant, W.C. *Handbook of North America Indians*, vol. 9. Washington, DC: Smithsonian Institution, 1979.

Marushima, H. *History of Japanese Stones*. Text in Japanese. Tokyo: Ishi-no-bi-sha Publishing Company, Ltd., 1992.

Matsuura, A., translated and adapted by Wil Lautenschlager. *An Introduction to Suiseki*. Tokyo: Otsukakogei-shinsha Co., Ltd., 2010

Mavor, J.W. Jr., and B. E. Dix. *Manitou*. Rochester, VT: Inner Traditions International, 1989.

Mowry, R. D. *Worlds Within Worlds: The Richard Rosenblum Collection of Chinese Scholar's Rocks*. Cambridge, MA: Harvard Art Museums, 1997.

Munroe, A. *The Third Mind: American Artists Contemplate Asia, 1860–1989*. New York: Guggenheim Museum Publications, 2009.

National Bonsai Foundation. *Awakening The Soul: The National Viewing Stone Collection of the National Bonsai and Penjing Museum at the U. S. National Arboretum*. Washington, DC: National Bonsai Foundation, 2000.

Noble, D.G. *An Archaeological Guide to Ancient Ruins of the Southwest*. Flagstaff, AZ: Northland Publishing, 2000.

Ojibwa. "Sacred Places in New England." www.nativeamericannetroots.net. Posted January 16, 2012; accessed February, 2014.

Ragle, N. *California Aiseki Kai Newsletter*. Laguna Beach, CA, monthly club newsletter.

Rivera, F.G. *Suiseki: The Japanese Art of Miniature Landscape Stones*. Berkeley, CA: Stone Bridge Press, 1977.

Squier, E.G. *Aboriginal Monuments of the State of New York*. Washington, DC: Smithsonian Contributions to Knowledge, vol. 2, 1849.

Tsuda, N. *A History of Japanese Art from Prehistory to the Taisho Period*. Tokyo, Rutland, Vermont & Singapore: Tuttle Publishing, 2009.

Tucker, M. *Suiseki & Viewing Stones. An American Perspective*. Flagstaff, AZ: Horizons West, 1996.

Tuohy, D.R., in Sturtevant, W.C., ed. *Handbook of North American Indians*, vol. 11. Washington, DC: Smithsonian Institution, 1986.

Whitley, D.S. *Cave Paintings and the Human Spirit*. Amherst, NY: Prometheus Books, 2009.

Wilson, W. *The History of Mineral Collecting*. 1530–1799. Tucson, AZ: The Mineralogical Record 25 (6): 1– 264, 1994.

Wu, Marshall P.S. *The Orchid Pavilion Gathering: Chinese Painting from the University of Michigan Museum of Art*. Ann Arbor, MI: Regents of the University of Michigan, 2000.

Xin, Yang, et al. *Three Thousand Years of Chinese Painting*; New Haven & London: Yale University Press, 2002.

Yeager, R. "Suiseki in Los Angeles," in *Waiting to be Discovered*. pp. 12–16. Summer, 1998.

A NOTE ON SELECTION

A panel of judges from diverse locations and backgrounds was designated to make the final selection of stones.

Rick Stiles hails from the Pacific Northwest, where he is an avid stone collector and author of numerous articles about North American stones. He founded the Pacific Northwest Bonsai Association stone collector's group in 2009.

Glen Reusch is a long-time stone collector and connoisseur from Virginia, who helped establish the Potomac Stone Club in the Washington, DC, area and the biennial International Stone Appreciation Symposium in Pennsylvania.

Richard Turner is an artist, curator, and Professor Emeritus, Department of Art, at Chapman University in Orange, CA, where he taught traditional and contemporary Asian art. His current studio work is inspired by his interest in stones.

Joe Grande, editor of *Bonsai & Stone Appreciation Magazine*, is a resident of Manitoba, Canada. He has been a principal in Granddesign, Ltd, a design studio operating in Winnipeg, Manitoba, for over forty years and was a lecturer for six years at the School of Art, University of Manitoba.

Thomas S. Elias conceived and executed this *North American Viewing Stones* publishing project. He is chairman of the Viewing Stone Association of North America and author of several books and numerous articles on Asian stone appreciation.

All judges are intimately acquainted with East Asian stone appreciation concepts, and embrace the idea of establishing a new framework for North American viewing stones. Based on the seven criteria for North American stones discussed earlier (pp. 8–9), a score of from 1 to 5 points was awarded for each, for a maximum of 35 points.

Except for Thomas S. Elias, who coordinated the project, judges did not have access to the names of owners or collectors. Instead, each stone was assigned a number. The panelists were provided only with the name of the stone, if one existed, the stone's size, and its geographic origin. For logistical reasons, panelists judged from photographs rather than by viewing actual stones. To ensure that no personal viewpoint dominated, no more than five stones were included from any single owner. In the end, from a total of 330 images submitted by sixty-three individuals and organizations submitted for possible inclusion, fewer than half, 151 stones, were selected.

A NOTE ON PRESENTATION

Stones are presented randomly rather than by geologic type, shape, or origin. Captions provided for each stone first give its name, typically evocative, indicating the feeling or image the stone conveys to its owner. Naming stones is a tradition inherited from East Asia and is important in helping novice collectors better appreciate them. If an owner has not provided one, the stone is given as unnamed. Next listed is where the stone was originally found. Some stones were legally collected in Death Valley, California, prior to its designation as a National Park. Today, collecting stones in national parks is not permitted.

Captions next provide a brief description highlighting features that make each an outstanding viewing stone. The mineral composition of the stone may be included here, although no attempt has been made to verify an owner's determination of this. Some stones have a single cut to accommodate a base. That information is provided when known. Next listed is the material used to make the base or support, and the artist who made it. If the name of the artist or the original collector is unknown, that is noted, although the current owner of the stone is always provided. Finally, dimensions of the stone are given in inches and converted to centimeters, rounded to the nearest tenth of a centimeter.

Although high score was the chief criteria for selection, diversity was deemed important, with a conscious effort made to represent a broad range of both sources and geological types. What might be seen as a lack of balance in geographical representation is due partly to the geology of North America, and also to the presence and activities of stone appreciation clubs and collectors. It is not surprising that just over half the stones included are from California, which has more active clubs and collectors than any other state. As well, the state's geological diversity—from ocean coastline, and coastal mountain ranges with active faulting, numerous river systems, and the extensive desert and semi-desert areas—provides excellent hunting grounds for quality stones.

In fact, the entire Pacific coastal region of North America is a prime area for finding outstanding viewing stones, as is the entire length of the Rocky Mountains. Stones from these regions combined account for perhaps eighty percent of the stones illustrated. The more ancient and weathered Appalachian Mountains have considerable potential, but river stones are not as readily available because so many are buried in sediments. Regrettably, the southern United States is not well represented, primarily due to a fewer people searching for quality stones. The extensive deposits of limestone in the South and windblown arid areas in the Southwest outside of southern California remain inadequately searched, and should one day yield impressive stones. We can take heart, however, that appreciation of viewing stones is still in its infancy in North America, and that numerous excellent stones will be found in new areas. We need only go out and search for them.

VIEWING STONES
OF NORTH AMERICA

1 (left) I **Grand Duchess**
Northern shore of Georgian Bay, Lake Huron, Ontario, Canada

This imposing vertically oriented figure stone can be viewed from two sides.

Base: Black Walnut, Tony Ankowicz I Collector: Tony Ankowicz I Owner: Tony Ankowicz
23 x 5 x 4 inches (58.4 x 12.7 x 10.2 cm)

2 | Planet Clare
Trinity River watershed, Northern California

Viewers using a bit of imagination can see North America, Europe, and Africa on this miniature Earth.

Base: Cliff Johnson I Collector: Unknown I Owner: Joseph Gaytan
7.5 x 6 x 4 inches (19 x 15.2 x 10.2 cm)

3 | Mountain of Flowers
Sierra River, Northern California

This large, egg-shaped stone has a surface pattern resembling a field of flowers.

Base: Painted basswood, Cliff Johnson | Collector: Ken McLeod | Owner: Cynthia McLeod
14 x 11 x 4 inches (35.6 x 28 x 10.2 cm)

4 (right) ǀ Majestic Mountain
Eel River, Northern California

This magnificent mountain composed of serpentine and jade was cut for placement in the base.

Base: Walnut, Joseph Gaytan ǀ Collector: Ken McLeod ǀ Owner: Ken McLeod
8 x 16 x 10 inches (20.3 x 40.6 x 25.4 cm)

5 ǀ Study in Movement
British Columbia, Canada

This Pacific Northwest coastal limestone suggests a runner, or a mythical creature preparing to take flight.

Base: Wood, Sean Smith ǀ Collector: Unknown ǀ Owner: Thomas S. Elias
6 x 5.8 x 4 inches (15.24 x 14.7 x 10.2 cm)

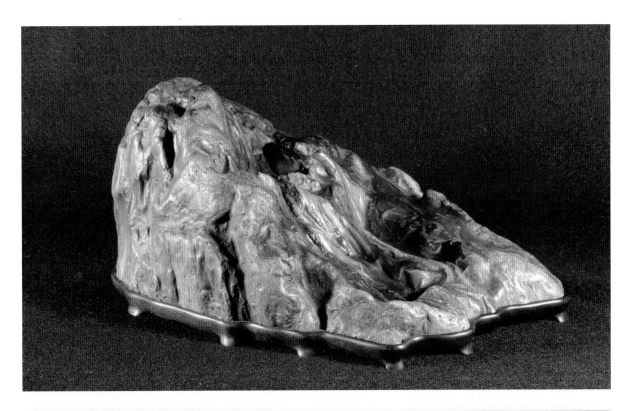

6 ⏐ Jade Mountain
Eel River watershed, Northern California

The mountain shape of this rare form of botryoidal jade has nice proportions complemented by the varying colors.

Base: Hardwood, Al Nelson ⏐ Collector: Ken McLeod ⏐ Owner: Ken McLeod
9 x 16 x 5 inches (22.9 x 40.6 x 12.7 cm)`

7 | Eagle
Northern California

The unusual shape of this natural piece of limestone and schist is reminiscent of a perching eagle.

Base: Walnut, Jerry Braswell | Collector: Ken McLeod | Owner: Cynthia McLeod
9 x 11 x 5 inches (23 x 28 x 12.7 cm)

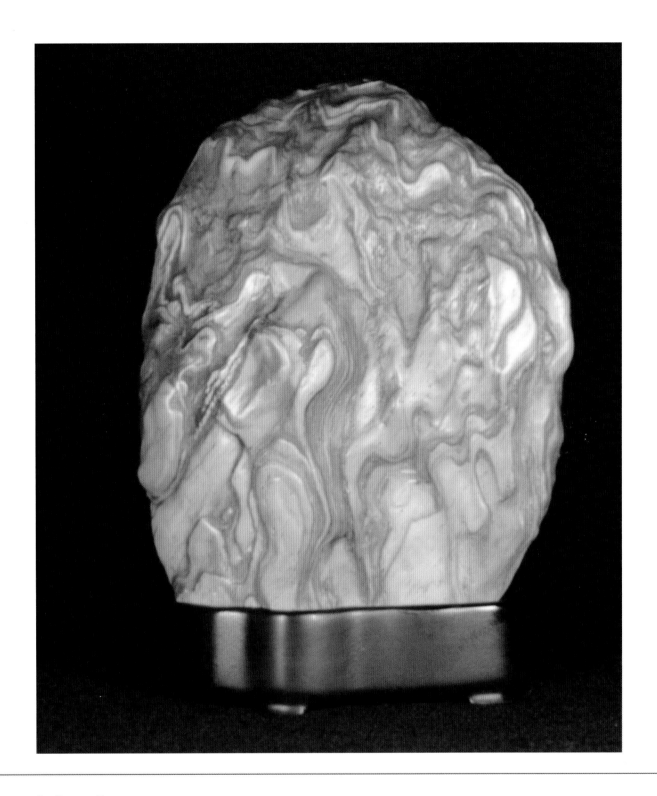

8 ı Dragon Face
Trinity River watershed, Northern California

This type of wavy serpentine with jade inclusions from this watershed was first introduced by Ken McLeod.

Base: Cherry wood, Jerry Braswell ı Collector: Ken McLeod ı Owner: Ken McLeod
12.5 x 8 x 4 inches (31.8 x 20.3 x 10.2 cm)

9 ı Old Pinnacle
Desert, Southern California

A tall, narrow piece of wind weathered quartz recalls the pointed peaks of Pinnacles National Park.

Base: Red sandalwood, Koji Suzuki ı Collector: Ralph Johnson ı Owner: Hiromi Nakaoji
6.3 x 3 x 2.3 inches (16 x 7.6 x 5.8 cm)

10 | Desert Mountain
Desert, Southern California

This limestone and quartz desert stone brings to mind a weathered mountain with canyons.

Base: Red sandalwood, Koji Suzuki | Collector: Ralph Johnson | Owner: Hiromi Nakaoji
3.5 x 8.5 x 3.6 inches (8.9 x 21.6 x 9.1 cm)

11 | Monterey Seacoast
Northern California

This dramatic combination of black basalt and light colored limestone may have been formed by strong waves crashing against the seacoast.

Base: Wood, Al Nelson | Collector: Ken McLeod | Owner: Cynthia McLeod
7 x 16 x 8 inches (17.8 x 40.6 x 20.3 cm)

12 (left) | **Ancient Sentinel**
San Bernardino County, Southern California

The original organic material of this section of tree limb has been entirely replaced with minerals, creating a heavily weathered stone exhibiting great strength and movement.

Base: Painted basswood, Cliff Johnson | Collector: Unknown | Owner: Thomas S. Elias
24 x 11 x 5.5 inches (61 x 28 x 14 cm)

13 | **Anticipation**
Yuha Desert, Southern California

This small ancient piece of folded granite and quartz resembles an insect or perhaps a frog ready to spring into action.

Base: Al Nelson | Collector: Al Nelson | Owner: Al Nelson
3 x 4 x 3 inches (7.6 x 10.2 x 7.6 cm)

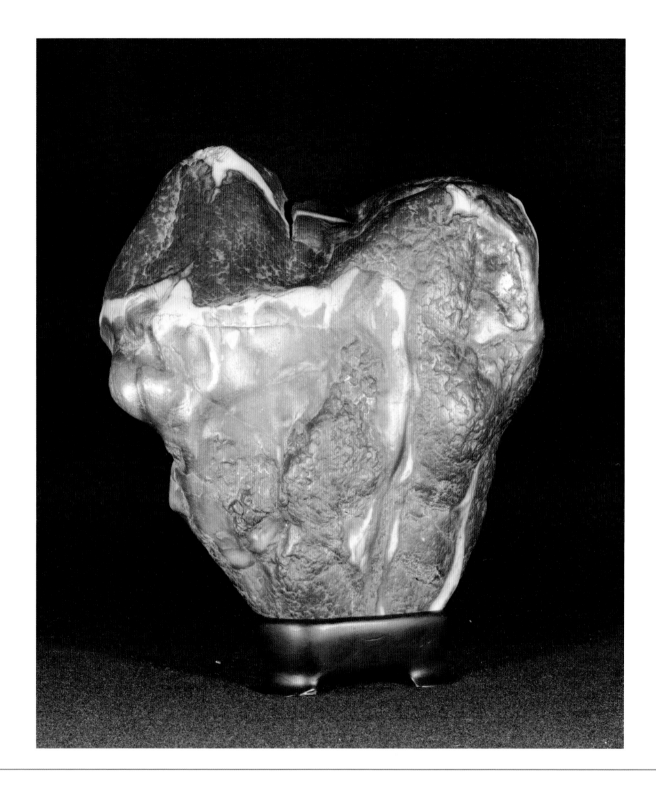

14 I My Heart
Trinity River Watershed, Northern California

A rare combination of jade, serpentine, and red jasper are deployed in an appealingly familiar shape.

Base: Cherry wood, Jerry Braswell I Collector: Ken McLeod I Owner: Ken McLeod
12 x 9 x 5 inches (30.5 x 22.9 x 12.7 cm)

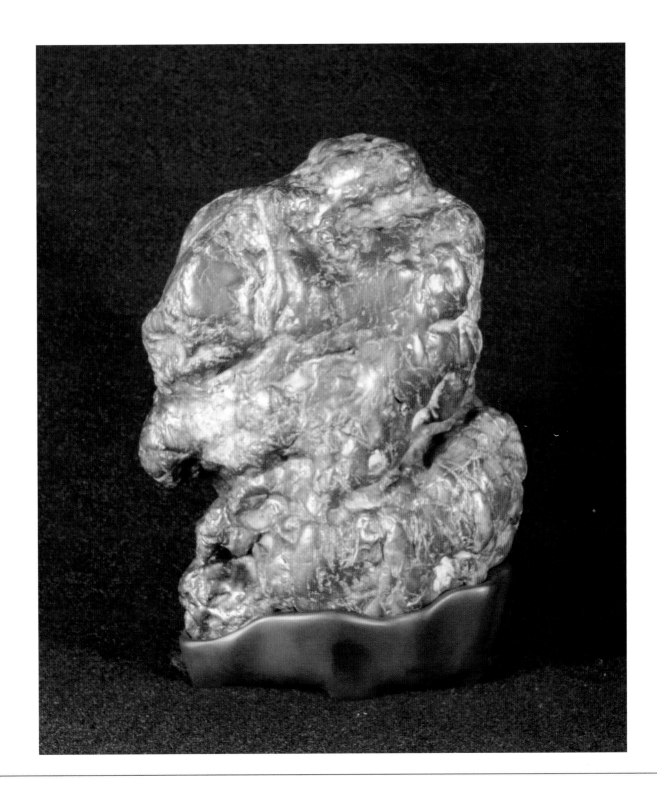

15 | Scary Face
Oregon

This type of stone, grossular garnet, is rare and seldom seen in collections.

Base: Hardwood, Al Nelson | Collector: Ken McLeod | Owner: Ken McLeod
9 x 7 x 3 inches (22.9 x 17.8 x 7.6 cm)

16 ı Waterfall in Evening
Trinity River, Northern California

This attractive stone of two-tone serpentine and possibly jade creates the illusion of a thin waterfall dropping from a steep vertical rock face.

Base: Walnut, Al Nelson ı Collector: Ken McLeod ı Owner: Al Nelson
11.5 x 4.5 x 2 inches (29.2 x 11.4 x 5 cm)

17 | Gentle Mountain
Pennsylvania

This mountain stone has been weathered down to low rounded peaks typical of the older Appalachian range of the Eastern United States, where it was found.

Base: hardwood, Sean Smith | Collector: Sean Smith | Owner: Gudrun Benz
2.6 x 8.3 x 4.7 inches (6.5 x 21 x 12 cm)

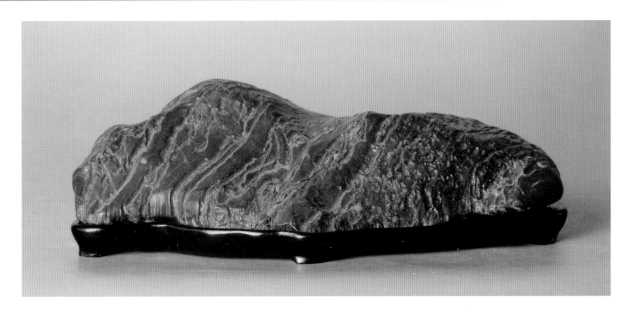

18 | Jade Mountain
Eel River, Northern California

This stunning piece of California jade appears naturally in a mountain shape.

Base: hardwood, Sean Smith | Collector: Unknown | Owner: Gudrun Benz
3 x 8.7 x 3.7 inches (7.5 x 22 x 9.5 cm)

19 | Teton Peaks

Lake Hill, Southern California

This black limestone was collected by one of the early leaders of North American stone appreciation, Melba Tucker. It has been partially polished.

Base: Ned Tucker | Collector: Melba Tucker | Owner: Dien Liang
6.3 x 8.8 x 6.7 inches (16 x 22.4 x 16.3 cm)

20 | American Gongshi

Mojave Desert, Southern California

Although this heavily weathered limestone resembles a large Chinese scholar's stone, this American desert piece is tiny.

Base: Wood, Cliff Johnson | Collector: Ann Horton | Owner: Ann Horton
3 x 1 x 1 inches (7.6 x 2.5 x 2.5 cm)

21 | Rising Emotions
Eel River, Northern California

This serpentine stone with soft undulating forms is found in northern California.

Base: Walnut, Paul Vasina | Collector: Jerry McNey | Owner: Al Nelson
7 x 3.5 x 2.5 inches (17.8 x 8.9 x 6.4 cm)

22 | Waterfall
Mohave Desert, Southern California

The quartz veins in this small stone give the appearance of a high mountain waterfall. This stone was first published in Melba Tucker's *Suiseki & Viewing Stones: An American Perspective* in 1996.

Base: Ned Tucker I Collector: Melba Tucker I Owner: Dien Liang
2.5 x 5 x 3 inches (6.4 x 12.7 x 7.6 cm)

23 (right) ⏐ Indian Blanket
Death Valley, Southern California

A striking desert scene is formed by the contrast between minerals comprising this stone, which was collected long before the area was designated a National Park.

Base: Painted basswood, Cliff Johnson ⏐ Collector: Ralph Johnson ⏐ Owner: Ralph Johnson
11 x 25 x 3 inches (28 x 63.5 x 7.6 cm)

24 ⏐ Twin Plateaus
Death Valley, Southern California

A combined granite and schist stone resembles twin mountain plateaus with several small lakes in the lowlands below.

Base: Cliff Johnson ⏐ Collector: Cliff Johnson ⏐ Owner: Ralph Johnson
9 x 20 x 13 inches (22.8 x 50.8 x 33 cm)

25 | Windswept Mountain
Palm Springs, Southern California

Over thousands of years this granite schist and quartz stone was scoured by sand-laden winds to form a miniature mountain. This stone was collected long before the area was designated a National Park.

Base: Cliff Johnson | Collector: Ralph Johnson | Owner: Ralph Johnson
9 x 17 x 12 inches (22.9 x 43.2 x 30.5 cm)

26 | Unnamed
Death Valley, Southern California

Rough mountain peaks stand out on this sloping limestone, collected in the early 1980s.

Base: Wood, artist unknown | Collector: Lee Roberts | Owner: Sean Horton
3.5 x 9 x 4 inches (8.9 x 22.9 x 10.2 cm)

27 | Mountain in Spring
Eel River, Northern California

This stone exudes the calmness of a gently sloping mountain scene.

Base: Japanese ceramic tray, artist unknown | Collector: Felix Rivera | Owner: Gudrun Benz
3.1 x 9.8 x 3.1 inches (8 x 25 x 8 cm)

28 | A Thousand Cranes
Northern California

An amazing display of these auspicious birds rising upward on a blustery day.

Base: Walnut, Peter Bloomer | Collector: Peter Bloomer | Owner: Peter Bloomer
8 x 6.5 x 4 inches (20.3 x 16.5 x 10.2 cm)

29 | Unnamed
Mojave Desert, Southern California

The shape and texture of this stone evoke an impressive ancient round-topped mountain.

Base: Wood, David Bennett | Collector: Unknown | Owner: William N. Valavanis
6.5 x 15 x 9 inches (16.6 x 38 x 23 cm)

30 (left) | **Sea Coast Tunnels**
Coastal British Columbia, Canada

An ancient weathered limestone resembles a coastal scene with hidden tunnels.

Base: Wood, Sean Smith | Collector: Unknown | Owner: Thomas S. Elias
4.2 x 8 x 4 inches (10.6 x 20.3 x 10.2 cm)

31 | **Shall We Dance?**
Murphy, California

This fascinating combination of carbonate stones gently undulate in unison, provoking a variety of imagery.

Base: Wood, Cliff Johnson | Collector: Hanne Povlson | Owner: Ralph Johnson
18 x 12 x 2 inches (45.7 x 30.5 x 5 cm)

32 I Turtle
Death Valley, Southern California

This weathered limestone with bits of quartz is a stark reminder that the desert regions of southern California once had many fresh water lakes filled with shelled creatures. This stone was collected long before the area was designated a National Park.

Base: Wood, Cliff Johnson I Collector: Ralph Johnson I Owner: Ralph Johnson
6 x 10 x 4 inches (15.2 x 25.4 x 10.2 cm)

33 | Mountain Retreat
Mojave Desert, Southern California

This strongly weathered combination of rhyolite and quartz creates an impression of sharply rising and imposing mountains.

Base: Black walnut, Cliff Johnson | Collector: Ann Horton | Owner: Sean Horton
6 x 11 x 4 inches (15.2 x 28 x 10.2 cm)

34 (right) ɪ Cascade
Mojave Desert, Southern California

Dynamism can be expressed by objects of varying sizes including this small limestone and quartz stone.

Base: Wood, Cliff Johnson ɪ Collector: Ann Horton ɪ Owner: Ann Horton
3.5 x 2 x 3 inches (8.9 x 5 x 7.6 cm)

35 ɪ Mountain Gods
Mojave Desert, Southern California

This small complex limestone has many rugged features for contemplation.

Base: Cliff Johnson ɪ Collector: Ann Horton ɪ Owner: Ann Horton
3 x 4 x 2 inches (7.6 x 10.2 x 5 cm)

36 | Living Arch
Mojave Desert, Southern California

The rugged texture and natural arch are part of the flowing form of this desert stone.

Base: Black walnut, Cliff Johnson | Collector: Ann Horton | Owner: Ann Horton
1.5 x 7 x 4 inches (3.8 x 17.8 x 10.2 cm)

37 (left) | Unnamed

Eel River, Northern California

This surface texture and color help create a mysterious but beautiful mountain.

Base: Walnut, Cliff Johnson | Collector: Cliff Johnson | Owner: Sean Horton
5 x 15 x 8 inches (12.7 x 38 x 20.3 cm)

38 | Earth in Motion

Mojave Desert, Southern California

The massive power of upward tectonic movement breaks and scatters the outer crust of the Earth.

Base: Black walnut, Cliff Johnson | Collector: Ann Horton | Owner: Ann Horton
5.5 x 9 x 5.5 inches (14 x 23 x 14 cm)

39 | Magnolia Blossom
Paterson, New Jersey

An impressive pattern stone presents a single, large Magnolia flower-like mineral deposit of pectolite in a basaltic matrix.

Base: Wood, E. Mark Rhyne | Collector: Unknown | Owner: Racie Rhyne
4.3 x 4 x 2.3 inches (11 x 10.2 x 5.8 cm)

40 | Unnamed
Trinity River Watershed, Northern California

The contrasting colors of this combination of jasper, serpentine, and jade make a beautiful stone.

Base: Wood, Joseph Gaytan | Collector: Ken McLeod | Owner: Freeman Wang
6 x 8.5 x 3 inches (15.2 x 21.6 x 7.6 cm)

41 (right) | Unnamed
Caledonia Mine, Upper Peninsula, Michigan

This piece of natural copper ore has the form of an exquisite natural Chinese scholar's rock.

Base: Cast bronze, Rick Stiles | Collector: Richard Whiteman | Owner: Rick Stiles
13.5 x 6 x 4 inches (35 x 15 x 10 cm)

42 | River Spirit
Stillaguamish River, Washington

This beautiful small stone suggests undulating wave formations and a flowing river.

Base: Cast bronze, Rick Stiles | Collector: Mimi Stiles | Owner: Mimi Stiles
2 x 5 x 5 inches (5 x 12.7 x 12.7 cm)

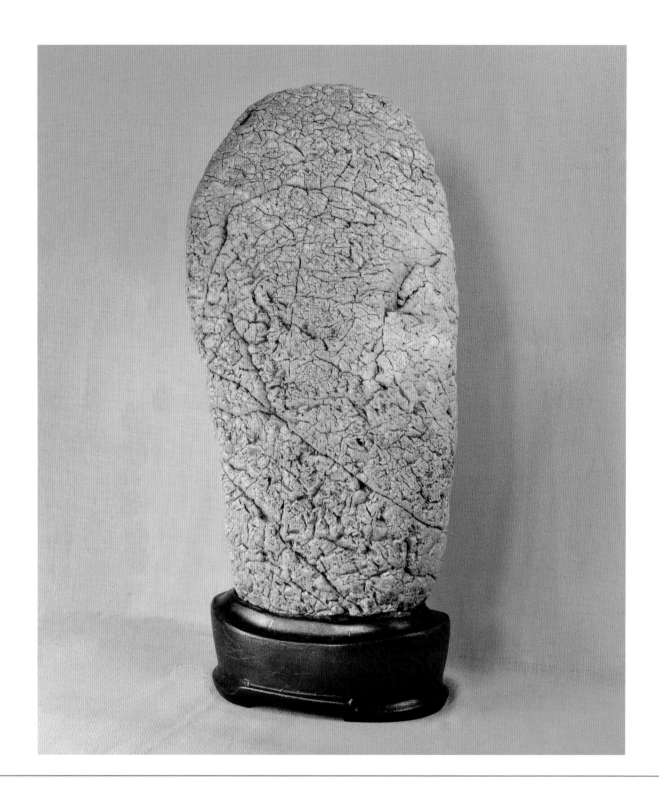

43 | Dragon Bone Stele
Wenatchee River, Washington

This magnificent upright stele with its water smoothed reticulated surface conveys an ancient character.

Base: Cast bronze, Rick Stiles | Collector: Mimi Stiles | Owner: Mimi Stiles
20 x 10 x 8 inches (51 x 25 x 20 cm)

44 | Unnamed
Stillaguamish River, Washington

The smooth surface and natural cleft evoke the rushing floodwaters of the Pacific Northwest.

Base: Bronze tray, Rick Stiles | Collector: Rick Stiles | Owner: Rick Stiles
5 x 5 x 4 inches (13 x 13 x 10 cm)

45 | Cascade High Ridge
Stillaguamish River, Washington

The shape of this glassy smooth mountain ridge and its river polished bronze color make an excellent viewing stone.

Base: Ceramic tray, Ron Lang | Collector: Mimi Stiles | Owner: Mimi Stiles
5 x 18 x 9 inches (13 x 46 x 25 cm)

46 (right) | Honorable New Man

Georgian Bay, Lake Huron, Ontario, Canada

This natural stone is reminiscent of a Giacometti sculpture, and pays homage to the Honorable Old Man, a stone from the Richard Rosenblum collection displayed at the Boston Museum of Fine Arts for many years.

Base: Cast bronze, Rick Stiles | Collector: Anthony Ankowicz | Owner: Rick Stiles
36 x 11 x 7 inches (92 x 28 x 17 cm)

47 | Hotei

Jade Cove, Big Sur, California

The beautiful shape, varied colors, and smooth surfaces of this natural botryoidal jade make an outstanding figure stone.

Base: Brazilian rosewood, Brent Wilson | Collector: Unknown | Owner: Sharon Somerfeld
5 x 3.4 x 3.8 inches (12.7 x 8.6 x 9.6 cm)

48 | Unnamed
Yuha Desert, Southern California

This desert ventifact has been shaped and polished by years of sand laden winds blowing against its greenish surface.

Base: Wood, Buzz Barry | Collector: Chris Barry | Owner: Buzz Barry
3 x 8.5 x 4 inches (7.6 x 21.6 x 10.2 cm)

49 ı Geisha
Mojave Desert, Southern California

Indian blanket stones have strongly contrasting colors and fascinating patterns, here possibly seen as a geisha or a tree.

Base: Wood, artist unknown ı Collector: Melba Tucker ı Owner: U.S. National Arboretum
9.5 x 9 x 7.5 inches (24 x 22.9 x 9.5 cm)

50 (left) ı Unnamed
Prince of Wales Island, Alaska

This naturally weathered beach stone offers an eccentric mountain peak with cloud features floating across its face.

Base: Walnut, Sean Smith ı Collector: Gary McWilliams ı Owner: Rick Stiles
4 x 11 x 4 inches (10.2 x 28 x 10.2 cm)

51 | The Wonder That Was India
Unknown

Each element of this sculptural ensemble brings its own associations. The prominent book title encourages us to think of the stone as a fragment of a ruined temple.

Base: Mixed media, polyester resin, wood, Richard Turner | Collector: Unknown | Owner: Richard Turner
15 x 11 x 6 inches (38 x 30 x 15.2 cm)

52 | Unnamed
Delaware County, Pennsylvania

The rugged appearance of this stone could well represent a mountain or island, and is exceptionally matched with the hardwood base.

Base: Wood, William Weber | Collector: Jim Hayes | Owner: U.S. National Arboretum
3.25 x 14 x 6.5 inches (8.3 x 35.6 x 16.5 cm)

53 | Unnamed
Eel River, Northern California

A beautiful twin-peaked island stone composed of serpentine is expertly matched with the bronze tray and sand depicting water.

Base: Bronze tray, artist unknown | Collector: Harry Hirao | Owner: U.S. National Arboretum
7 x 18.5 x 9.8 inches (17.8 x 47 x 25 cm)

54 ı Sitting Bison
Eel River, Northern California

This piece of natural nephrite and serpentine has beautiful colors with its greens, gold, and cream-colored striations.

Base: Walnut, Brent Wilson ı Collector: Brent Wilson ı Owner: Brent Wilson
12 x 17 x 10 inches (30.5 x 43.2 x 25.4 cm)

55 | Building Supply
Palm Springs, California

A broken piece of limestone supported on a base made from scraps of wood, echoes the strata of the stone and references the materials used in home construction.

Base: Wood, filler, urethane, Richard Turner I Collector: Richard Turner I Owner: Richard Turner
15 x 11 x 6 inches (38 x 28 x 15.2 cm)

56 | Unnamed

Eel River, Northern California

The steeply rising mountain stone is reminiscence of peaks seen in the Sierra Mountains in California.

Base: Wood, Cliff Johnson | Collector: Richard Manning | Owner: U.S. National Arboretum
8.5 x 21.5 x 12.75 inches (22 x 54.6 x 32.4 cm)

57 | Dark Shore

Unknown

A black base suggests the roiling night sea crashing against the base of steep cliffs along the shore.

Base: Painted plywood and polyester resin, Richard Turner | Collector: Unknown | Owner: Richard Turner
4 x 21 x 6 inches (10.2 x 53.3 x 15.2 cm)

58 | Mountain Sprite Dancing
Taylor River Valley, Rocky Mountains, Colorado

This limestone fragment shaped underground by acid water leaching gives the appearance of a graceful dancer with head bowed.

Base: Black walnut, Allan Hills | Collector: Allan Hills | Owner: Allan Hills
7.5 x 5.75 x 3.75 inches (19 x 14.7 x 9.5 cm)

59 (right) ı Glaciated Mountains of Patagonia
Taylor River Valley, Rocky Mountains, Colorado

This stone reminds its owner of the majestic peaks of Cuernos del Paine in Torres del Paine National Park, southern Chile, where glaciers cut deep and narrow U-shaped gorges in granite, leaving flat-topped horns capped by layers of metamorphic rock.

Base: Walnut, Allan Hills ı Collector: Allan Hills ı Owner: Allan Hills
7.5 x 13.8 x 7 inches (19 x 35 x 18 cm)

60 ı Water Poem
Cache la Poudre River, Rocky Mountains, Colorado

The four lobes make a visually dynamic stone that was created by rapidly flowing water over thousands of years.

Base: Mahogany, Darrell Whitley ı Collector: Darrell Whitley ı Owner: Darrell Whitley
12.3 x 16.5 x 9 inches (26 x 42 x 23 cm)

61 ı Echo of the Great Wave

Taylor River Valley, Rocky Mountains, Colorado

A residual fragment of dark limestone, shaped underground by acid ground water leaching and set in a tray with sand, has the appearance of a giant wave.

Base: Rakuware tray, Kendall Coniff ı Collector: Allan Hills ı Owner: Allan Hills
6 x 8.3 x 6.5 inches (15 x 21 x 16.5 cm)

62 ┃ Glacial Canyons
Snowy Range, Wyoming

This stone has the appearance of a glacial carved high mountain landscape with deep canyons.

Base: Cherry, Paul Gilbert ┃ Collector: Paul Gilbert ┃ Owner: Paul Gilbert
17 x 8 x 4 inches (43 x 20 x 10 cm)

63 (left) ⏐ Unnamed
Stillaguamish River, Washington

An imposing high plateau serpentine stone with appealing color.

Base: Polyester resin ⏐ Collector: Peter Bloomer ⏐ Owner: U.S. National Arboretum
6 x 12.3 x 5.3 inches (15.2 x 31.2 x 13.5 cm)

64 ⏐ Arch at Rattlesnake Canyon
South Park, Colorado

This stone depicts the beauty and grace of the natural arches that are iconic elements of North American West.

Base: Walnut, Larry Jackel ⏐ Collector: Larry Jackel ⏐ Owner: Larry Jackel
5.5 x 8 x 3.5 inches (14 x 20 x 9 cm)

65 | Eskimo Girl

Prince of Wales Island, Alaska

This conglomerate of volcanic rocks and limestone gives the appearance of a playful Eskimo girl in a hooded parka.

Base: Walnut, Rick Klauber | Collector: Gary McWilliams | Owner: Gary McWilliams
14.5 x 9.8 x 5.5 inches (36.8 x 24.9 x 14 cm)

66 | Swirls
Prince of Wales island, Alaska

This abstract object stone of black marble is eroded in an eye-catching pattern suggesting a wave.

Base: Wood slab | Collector: Gary McWilliams | Owner: Gary McWilliams
5 x 11.2 x 5 inches (12.7 x 28.4 x 12.7 cm)

67 | Infinite Mountain Range
Clear Creek, San Benito County, California

This deep black, smooth, satin-like surface and shape of this stone can be viewed as a distant mountain range or perhaps an island.

Base: Cherry | Collector: Hans Thern | Owner: Robert McKenzie
2.5 x 10 x 1.5 inches (6.4 x 25.4 x 3.8 cm)

68 (right) ı Delicate Spire
Vancouver Island, Canada

This delicate irregular piece of weathered limestone resembles a spire or perhaps a traditional Chinese stone.

Base: Walnut, Larry Jackel ı Collector: Larry Jackel ı Owner: Larry Jackel
4.5 x 1.5 x 1.5 inches (11.5 x 4 x 4 cm)

69 ı **Turquoise Canyon Falls**
Dry Creek, Mendocino County, California

A near perfect waterfall cascades down a dark, turquoise colored mountain.

Base: Donald Dupras ı Collector: Donald Dupras ı Owner: Donald Dupras
3 x 6 x 3 inches (7.6 x 15.2 x 7.6 cm)

70 (right) | **The Wall at Horse Thief Canyon**

Snowy Range, Wyoming

It is easy to image this stone as a massive, solid limestone wall of a hidden canyon.

Base: Cherry, Larry Jackel | Collector: Paul Gilbert | Owner: Larry Jackel
6 x 17 x 7 inches (13 x 43 x 18 cm)

71 | **Bursting Out**

Eel River, Northern California

This stone has beautiful hues of pink and green as if these rich colors are seeking to burst through the deep patina.

Base: Mahogany, Paul Gilbert | Collector: Ken McLeod | Owner: Theresa and Bill Hertneky
4.7 x 7 x 1.6 inches (12 x 18 x 4 cm)

72 | Spring Rain
Taylor Reservoir, Colorado

A beautiful mountain stone features lovely peaks and a valley between, with white veins resembling the cold spring runoff of melting snow.

Base: Cherry, Paul Gilbert I Collector: Paul Gilbert I Owner: Paul Gilbert
15 x 7 x 6.5 inches (39 x 18 x 17 cm)

73 | Field of Peonies
Colorado River, Colorado

This lovely porphyry flower stone has a natural sheen and patina.

Base: Rosewood, Paul Gilbert | Collector: Paul Gilbert | Owner: Paul Gilbert
12 x 12 x 5 inches (30 x 30 x 13 cm)

74 ⅼ Unnamed
Salt Lake County, Rocky Mountains, Utah

This strongly eroded limestone is typical of the weathered rock formations of the American Southwest. The deep ledges and shelves strengthen an upward movement.

Base: Walnut, Darrell Whitley ⅼ Collector: Darrell Whitley ⅼ Owner: Darrell Whitley
6 x 5 x 3.5 inches (15.2 x 12.7 x 8.9 cm)

75 | Tunnel of a Lifetime

Stillaguamish River, Washington

The many ledges, pits, and flat floor entice viewers to enter the tunnel.

Base: Cherry, Rick Klauber | Collector: Rick Klauber | Owner: Rick Klauber
6 x 18 x 7 inches (15.2 x 45.7 x 17.8 cm)

76 | Snowy Mountain with Stream

Prince of Wales Island, Alaska

A mountain stone with a striking resemblance to Mt. Rainier in Washington state, with its permanent snow fields and tumbling streams.

Base: Elm, Rick Klauber | Collector: Gary McWilliams | Owner: Gary McWilliams
11.5 x 8.5 x 5 inches (29.2 x 21.6 x 12.7 cm)

77 (left) | Moonlit Fuji

Clear Creek, Northern California

The extraordinary piece of serpentine is a miniature version of the famous Mount Fuji.

Base: Walnut, Mas Nakajima | Collector: Mas Nakajima | Owner: Mas Nakajima
12 x 7 x 6 inches (30.5 x 17.8 x 15.2 cm)

78 (right) | Cave Mountain
Northern California

The excellent shape, multi-colored stone with a pleasing balance of textures makes a beautiful viewing stone. A single baseline cut was made to this stone.

Base: Walnut, William B. Meran | Collector: William B. Meran | Owner: William B. Meran
4 x 10.5 x 6.3 inches (10.2 x 26.7 x 16 cm)

79 | Windswept Shore Arch
Prince of Wales Island, Southeastern Alaska

An arch like this might be found in a costal scene from Monterey north to British Columbia.

Base: Walnut, Rick Klauber | Collector: Gary McWilliams | Owner: Rick Klauber
5.3 x 6.5 x 2.5 inches (13.5 x 16.5 x 6.4 cm)

80 I **Resting Bison**
Hampton Buttes, Oregon

A bison rests on the warm ground near one of the bubbling pools of Yellowstone National Park.

Base: Ceramic tray with sand I Collector: Rick Klauber I Owner: Rick Klauber
6 x 9.5 x 3.8 inches (15.2 x 24.1 x 9.5 cm)

81 | Autumn on the Mountain
Eel River, Northern California

This stone appears to some as the multi-colored pinnacles of a mountain, to others as a person wrapped in a colored blanket.

Base: Wood, Al Nelson | Collector: Cliff Johnson | Owner: Joseph Gaytan
18 x 10 x 6 inches (45.7 x 25.4 x 15.2 cm)

82 | Unnamed
Rappahannock River, Virginia

The gently sloping lines of the left side contrast with the steep cliff on the right side of this plateau stone.

Base: Walnut, Sean Smith | Collector: Jack Sustic | Owner: Jack Sustic
3.2 x 9.5 x 7 inches (8.1 x 24 x 17.8 cm)

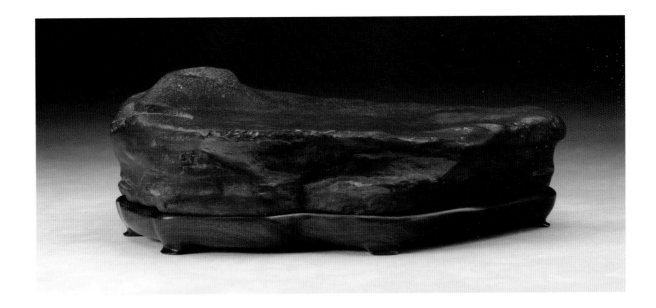

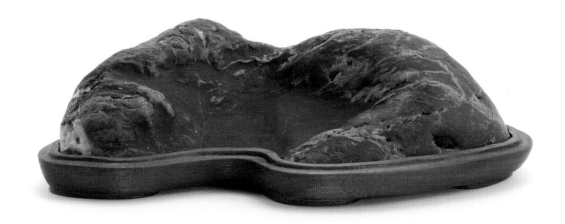

83 | Blue Mountain
Northern California

Its shape and the mix of blue and green colors make this an appealing stone. It has a cut base.

Base: Walnut, William B. Meran | Collector: Diane Meran | Owner: William B. Meran
2.5 x 8.7 x 5.5 inches (6.4 x 22 x 14 cm)

84 (left) I **Tsunami**

Cedar River, Washington

A piece of fossil wood appears like a giant wave about to crash against the land.

Base: Walnut, Patrick Metiva I Collector: Patrick Metiva I Owner: Patrick Metiva
8 x 4.4 x 3.4 inches (20.3 x 11.2 x 8.6 cm)

85 I **Flat Iron Mountain**

Sawmill Creek, San Benito County, California

This stone evokes a distant view of the Matterhorn in the summertime.

Base: Rosewood, Donald Dupras I Collector: Hans Thern I Owner: Donald Dupras
3.5 x 5.5 x 4 inches (8.9 x 14 x 10.2 cm)

86 | Japanese Cranes

Prince of Wales Island, Alaska

This pattern stone appears as a series of rapid photographs of cranes landing at the edge of a pool.

Base: None | Collector: Gary McWilliams | Owner: Gary McWilliams
10.5 x 7 x 4 inches (26.7 x 17.8 x 10.2 cm)

87 | Klamath Forest
Black Butte Lake, Northern California

The shape and texture of this gently sloping graceful stone epitomizes the Japanese concept of *wabi*, or subdued beauty.

Base: Walnut, Mas Nakajima | Collector: Janet Roth | Owner: Janet Roth
5 x 11.5 x 4.5 inches (12.7 x 29.2 x 11.4 cm)

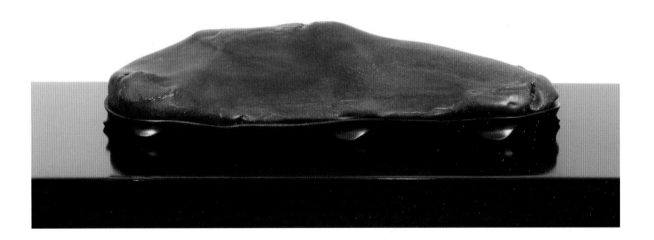

88 | Unnamed
Eel River, Northern California

A distant mountain stone that suggest an endless view.

Base: Walnut, Cliff Johnson | Collector: Larry Ragle | Owner: Larry Ragle
3.5 x 16 x 6 inches (8.9 x 40.6 x 15.2 cm)

89 | Three Peak Happiness
Eel River, Northern California

The combination of rough and smooth surfaces add to this beautiful mountain stone. A single baseline cut was made.

Base: Cherry, Robert McKenzie I Collector: Robert McKenzie I Owner: Robert McKenzie
4 x 14 x 4 inches (10.2 x 35.6 x 10.2 cm)

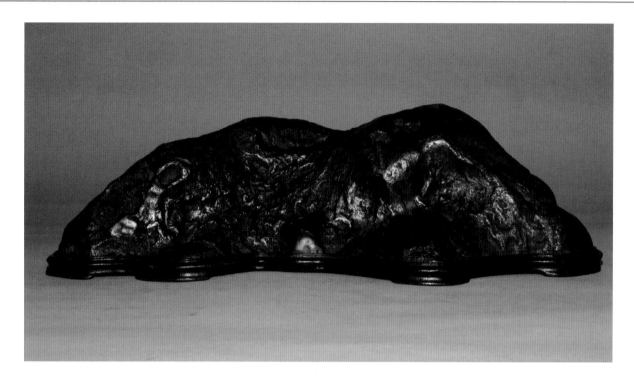

90 | Unnamed
Eel River, Northern California

This figure stone projects an image of a buffalo, native to the western United States.

Base: Walnut, Larry Ragle | Collector: Larry Ragle | Owner: Larry Ragle
9 x 11 x 6 inches (22.9 x 28 x 15.2 cm)

91 (left) | Last Rays of Sunset
Southern Oregon

This single peak, multi-colored mountain stone exhibits alternating patches of light and dark with orange highlights suggesting the reflection of the setting sun. A single baseline cut was made to this stone.

Base: Walnut, Robert McKenzie | Collector: Steve Beck | Owner: Robert McKenzie
4 x 10 x 4 inches (10.2 x 25.4 x 10.2 cm)

92 | Unnamed
Eel River, Northern California

This serpentine boulder suggest a Northern California coastal scene.

Base: Polyester resin, Larry Ragle | Collector: Nina Ragle | Owner: Nina Ragle
11.5 x 14 x 9 inches (29.2 x 35.6 x 22.9 cm)

93 (right) | First Snow
Clear Creek, Northern California

This appears as a powerful and stately mountain with a light dusting of snow at the peak, perhaps the first snow of late autumn.

Base: Walnut, Mas Nakajima | Collector: Mas Nakajima | Owner: Mas Nakajima
6 x 16 x 7 inches (15.2 x 40.6 x 17.8 cm)

94 | Unnamed

Eel River, Northern California

This gently sloping mountain stone has an elegant yet peaceful appearance.

Base: Walnut, Gail Middleton | Collector: Larry Ragle | Owner: Larry Ragle
5.5 x 14 x 9 inches (14 x 35.6 x 22.9 cm)

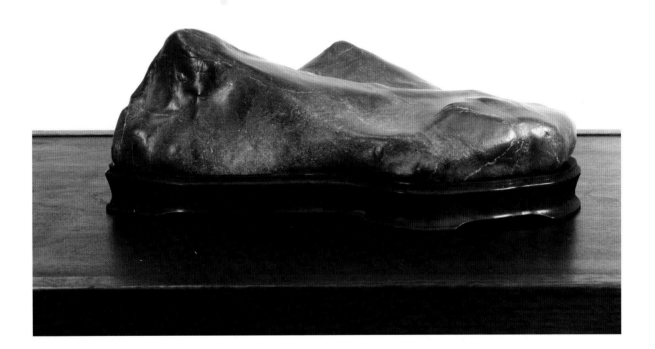

95 I Soaring Peak
Stoney Ford, Northern California

The well-balanced shape to this single peak mountain stone is complemented by the smooth, naturally river polished green jasper.

Base: Cherry, Robert McKenzie I Collector: Robert McKenzie I Owner: Robert McKenzie
3.5 x 9 x 5 inches (8.9 x 22.9 x 12.7 cm)

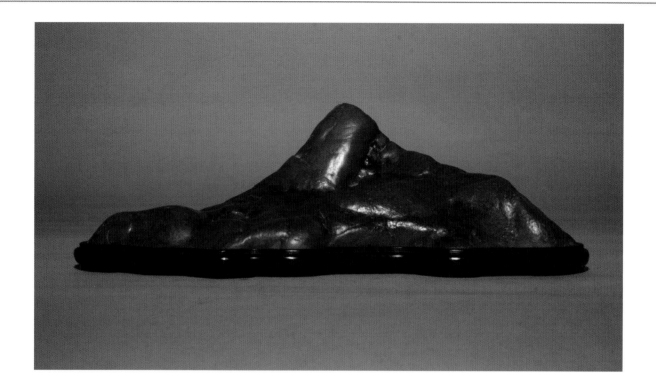

96 | Waiting for Salmon

Cascade Mountains, Washington

An Alaskan bear waits in a river to catch migrating salmon.

Base: Wood, artist unknown | Collector: Joel Schwarz | Owner: Joel Schwarz
10.5 x 8 x 7.5 inches (26.7 x 20.3 x 19 cm)

97 (left) | Beach Arch

Sauk River, Washington

This small stone resembles a coastal arch with the lines of the stone base forming waves, the colors of the base complementing those of the stone.

Base: Sandstone, Patrick Metiva | Collector: Patrick Metiva | Owner: Patrick Metiva
2.2 x 3.5 x 1.1 inches (5.6 x 8.9 x 2.8 cm)

98 I Pond Prince

Sauk River, Washington

This object stone, a green serpentine with wavy surface patterns, makes a convincing bullfrog, and is beautifully matched with its lily pad-like base.

Base: Poplar, Edd Kuehn I Collector: Edd Keuhn I Owner: Edd Kuehn
7 x 6.5 x 6 inches (17.8 x 16.5 x 15.2 cm)

99 | Western Plateau
Umpqua River, Oregon

A marvelous example of a classic plateau-shaped stone with deep rich color and excellent patina, matched perfectly with skillfully carved base.

Base: Wood, Sean Smith | Collector: Martin Schmalenberg | Owner: Martin Schmalenberg
3 x 9.5 x 5.5 inches (7.6 x 24 x 14 cm)

100 | Unnamed
Van Deusen River, Northern California

It is easy to image this as a rugged desert mountain with bold red colors and strong texture.

Base: Wood, Sean Smith | Collector: Martin Schmalenberg | Owner: Martin Schmalenberg
4.5 x 16 x 5 inches (11.4 x 40.6 x 12.7 cm)

101 (right) | Doppler Dazzler
San Benito County, California

The spirit of this stone is buoyant enough to evoke a fluffy cloud floating by.

Base: Walnut, Edd Kuehn | Collector: Edd Kuehn | Owner: Edd Kuehn
2.8 x 5 x 4.5 inches (7.1 x 12.7 x 11.4 cm)

102 | Unnamed
California

The colorful lines and patterns of this object stone give it movement, evoking a bird ready to take flight, perhaps in search of a mate?

Base: Carved mopani branch, Edd Kuehn | Collector: Edd Kuehn | Owner: Edd Kuehn
4 x 7 x 3.5 inches (10.2 x 17.8 x 8.9 cm)

103 (right) | Island Retreat
Eel River, Northern California

A small cove lies at the base of the steeply sloping island stone. Serpentine stones such as this make excellent viewing stones.

Base: Ceramic tray, Tokoname, Japan | Collector: Ken McLeod | Owner: Lindsay Bebb
5.1 x 8.7 x 5.5 inches (13 x 22 x 14 cm)

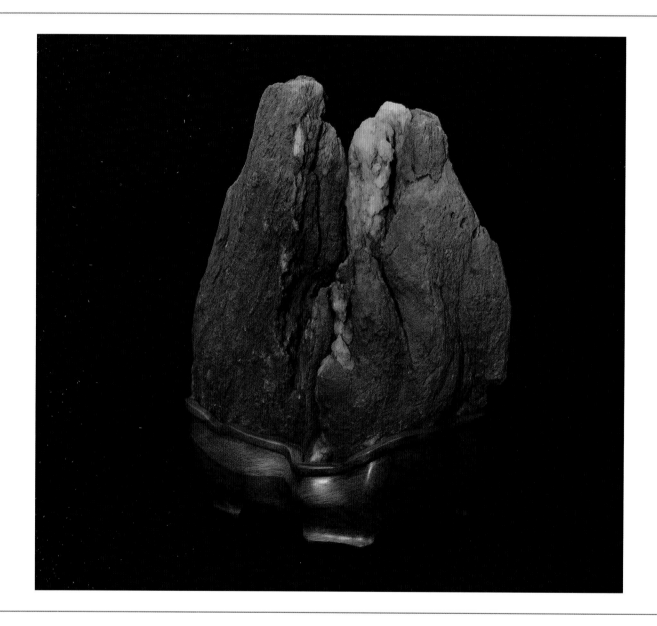

104 | Uncle
Northern California

This mountain stone with two sharply rising twin peaks resembles aspects of the North Cascade Mountains of Washington.

Base: Walnut, Patrick Metiva | Collector: Hirosi Kofu | Owner: Edd Kuehn
8.5 x 7 x 5 inches (21.6 x 17.8 x 12.7 cm)

105 | Unnamed
Chesapeake Bay, Maryland

This low, gently sloping mountain stone with its deep rich reddish brown color rests on a perfectly matched wooden base.

Base: Wood, Brian McCarthy | Collector: Brian McCarthy | Owner: Brian McCarthy
2 x 8 x 4.5 inches (5 x 20.3 x 11.4 cm)

106 (right) I Slope Stone
Southeastern Pennsylvania

This gently sloping stone with its subtle colors is an understated beauty.

Base: Mahogany, Brian McCarthy I Collector: Brian McCarthy I Owner: Brian McCarthy
3 x 12 x 5 inches (7.6 x 30.5 x 12.7 cm)

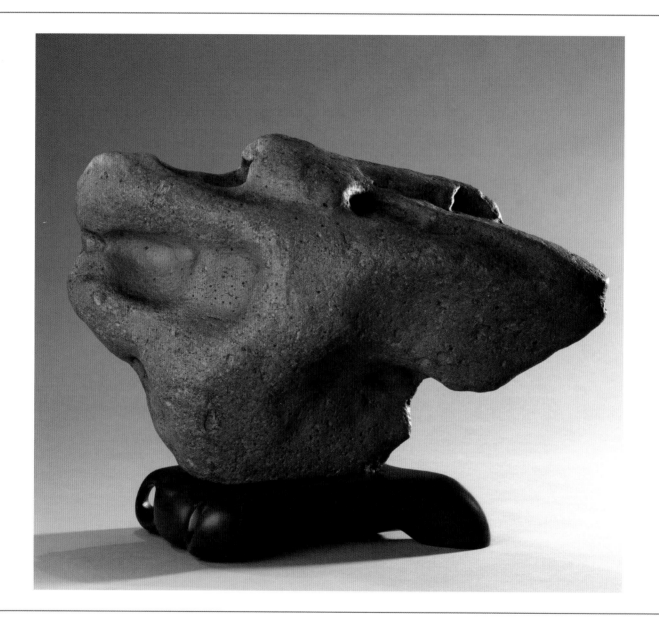

107 I Dragon Head
Deer Creek, Maryland

The physical features of this large stone resemble a dragon's head, albeit a benign Eastern dragon.

Base: Cherry, Brian McCarthy I Collector: Brian McCarthy I Owner: Brian McCarthy
16 x 23 x 10 inches (40.6 x 58.4 x 25.4 cm)

108 ı Unnamed

Patapsco River, Maryland

The shape and proportions of this intense metallic black stone make an idea coastal rock.

Base: Walnut, Brian McCarthy ı Collector: Brian McCarthy ı Owner: Brian McCarthy
3 x 10 x 7 inches (6.6 x 25.4 x 17.8 cm)

109 | Unnamed

Eel River, Northern California

This rugged yet beautiful mountain stone has traces of red jasper mirroring sunlight and moss to give the illusion of forest on the slopes. The moss is a departure from the traditional Japanese way of displaying a stone.

Base: Bronze tray, Fujisawa Roseki | Collector: Mas Nakajima | Owner: Sam Edge
4.5 x 15.5 x 5 inches (11.4 x 39.4 x 12.7 cm)

110 | Mysterious Stone

Eel River, Northern California

This serpentine with jasper inclusion appears like a rugged mountain peak rising out of the sea.

Base: Antique bronze tray | Collector: Martin Schmalenberg | Owner: Martin Schmalenberg
6 x 13 x 6 inches (15.2 x 33 x 15.2 cm)

111 ⎪ "... and then there was light"
Creek, Northern California

A view of the creation shows turbulence and initial flashes of light.

Base: Walnut, Sean Smith ⎪ Collector: Glen Reusch ⎪ Owner: Glen Reusch
9 x 14 x 8 inches (22.9 x 35.6 x 20.3 cm)

112 (left) | The Little Refugee
Mores Creek, Idaho

This basaltic figure stone resembles a little girl carrying a doll and burdened with a load on her back.

Base: Walnut, Sean Smith | Collector: Glen Reusch | Owner: Glen Reusch
6.5 x 8 x 4 inches (16.5 x 20.3 x 10.2 cm)

113 | Unnamed
Mores Creek, Idaho

This volcanic rock appears as an imaginary mountain with a dramatic valley made of the large vesicles.

Base: Walnut, Sean Smith | Collector: Glen Reusch | Owner: Glen Reusch
9.5 x 5 x 6 inches (24.1 x 12.7 x 15.2 cm)

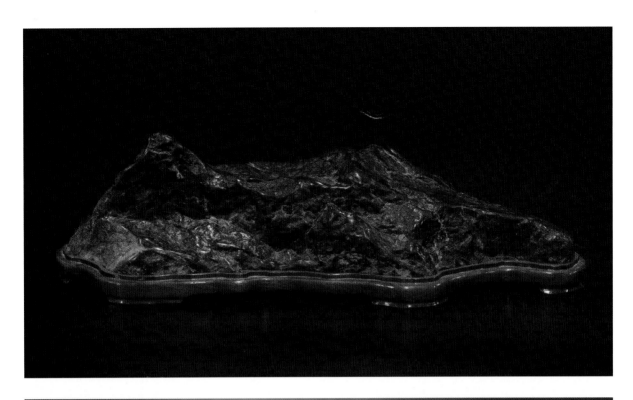

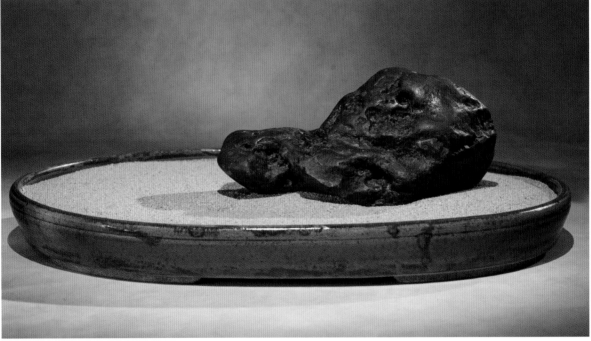

114 ı Unnamed
Shenandoah River, Virginia

A coastal rock shows the effects of heavy wave actions on its flanks.

Base: Ceramic tray, Nick Lenz ı Collector: Glen Reusch ı Owner: Glen Reusch
9 x 5.5 x 4.5 inches (22.9 x 14 x 11.4 cm)

115 (left) | Unnamed
Pennsylvania

The fine black color and patina help make this an excellent distant mountain stone.

Base: Walnut, Sean Smith | Collector: Sean Smith | Owner: Robert Blankfield
3.5 x 12.2 x 5.2 inches (8.9 x 31 x 13.2 cm)

116 | Unnamed
New York State

This sheer mountain peak thrusting through a broken ring of clouds, was collected by one of the great proponents of North American stone appreciation.

Base: Walnut, Robert Blankfield | Collector: Yuji Yoshimura | Owner: Dawn Blankfield
9.5 x 7 x 7.5 inches (24.1 x 17.8 x 19 cm)

117 (right) ı **Yellow Crane Peak**
James River, Virginia

This abstract shape and both smooth and rough textures of this stone suggest a powerful yet graceful human form or a soaring landscape peak.

Base: Wood root, Chris Cochrane ı Collector: Chris Cochrane ı Owner: Chris Cochrane
10.5 x 7.3 x 4.5 inches (26.7 x 18.5 x 11.4 cm)

118 ı **The Dance**
Shenandoah Valley, Boyce, Virginia

The abstract but recognizable patterns suggest joyful, spontaneous, and rhythmic dance movements.

Base: Cherry, Chris Cochrane ı Collector: Chris Cochrane ı Owner: Chris Cochrane
11.2 x 10 x 5.8 inches (28.4 x 25 x 14.7 cm)

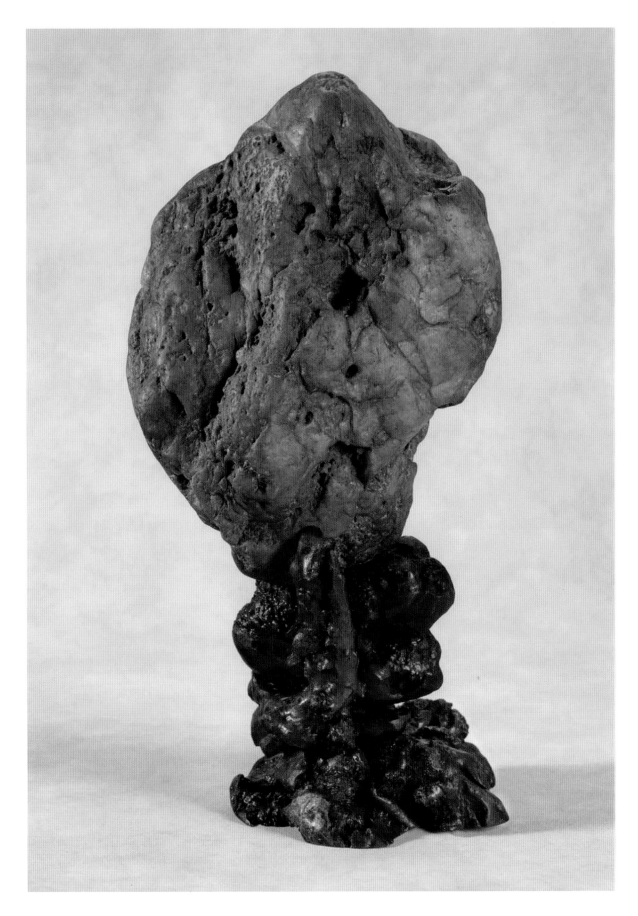

119 | Twin Peaks
Shenandoah River, Virginia

This was one stone that was cut in half to form the twin peaks of a mountain range.

Base: Mahogany, Michael J. Collella | Collector: Michael J. Colella | Owner: Michael J. Colella
3 x 6 x 5 inches (7.6 x 15.2 x 12.7 cm)

120 (left) I **Hut with Broken Roof**
James River, Virginia

This small hut-shaped stone at night may be reflecting a star-filled sky framed by broken reeds.

Base: East Indian rosewood I Collector: Chris Cochrane I Owner: Chris Cochrane
3.5 x 5.5 x 2.8 inches (8.9 x 14 x 7.1 cm)

121 I **Alone on an Island**
Location unknown

This iron stained peak sits on its own island in the midst of pure white sand from Mexico.

Base: Ebony with sand, Michael J. Colella I Collector: Michael J. Colella I Owner: Michael J. Colella
4 x 3 x 3 inches (10.2 x 7.6 x 7.6 cm)

122 I Unnamed

Desert, Southern California

This severely wind and sand sculpted desert stone forms a ledge to escape the sun.

Base: Wood, Cliff Johnson I Collector: Hanne Povslen I Owner: Hanne Povlsen
5 x 15 x 11 inches (12.7 x 38 x 28 cm)

123 | Unnamed
Unknown

This beautiful and elegant concretion was collected by a North American stone hunter long before Asian stone appreciation concepts were known in North America.

Base: Cocobolo, Tony Ankowicz | Collector: Bryon Buckeridge | Owner: Tony Ankowicz
3 x 3 x 3 inches (7.6 x 7.6 x 7.6 cm)

124 | Unnamed
Desert, Southern California

This rugged, deeply weathered limestone with quartz deposits on many edges gives the impression of young mountains.

Base: Wood, Cliff Johnson | Collector: Hanne Povlsen | Owner: Hanne Povlsen
3.5 x 15 x 3 inches (8.9 x 38 x 7.6 cm)

125 | Unnamed
Desert, Southern California

The sharp peaks of the wind weathered piece of rhyolite make an imposing distant mountain stone.

Base: Wood, Cliff Johnson | Collector: Hanne Povlsen | Owner: Hanne Povlsen
3 x 13.5 x 3.5 inches (7.6 x 34.3 x 8.9 cm)

126 | Layers
Quarry, Western Maryland

The interesting shape and beautifully contrasting light and gray layers make a nice viewing stone.

Base: Ebony, Michael J. Colella | Collector: Michael J. Colella | Owner: Michael J. Colella
9 x 9 x 2 inches (22.9 x 22.9 x 5 cm)

127 (right) ı Unnamed
Desert, Southern California

An imposing rugged piece of multi-colored rhyolite formed by sand-laden winds in a hot desert environment.

Base: Wood, Cliff Johnson ı Collector: Hanne Povlsen ı Owner: Hanne Povlsen
5 x 10 x 7 inches (12.7 x 25.4 x 17.8 cm)

128 ı Unnamed
Desert, Southern California

This layered rhyolite stone has been strongly eroded by sand-laden winds.

Base: Wood, Cliff Johnson ı Collector: Hanne Povlsen ı Owner: Hanne Povlsen
5 x 5.5 x 4 inches (12.7 x 14 x 10.2 cm)

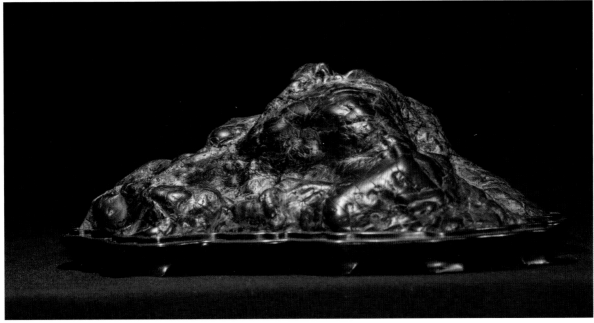

129 ı Unnamed

Van Deusen River, Northern California

This dense swirling serpentine with its deep rich color and wonderful patina makes an excellent mountain stone. The lowest part of the stone has been cut to facilitate a base.

Base: Wood, Sean Smith ı Collector: Martin Schmalenberg ı Owner: Martin Schmalenberg
5 x 13 x 10 inches (12.7 x 33 x 25.4 cm)

130 ı Cotton Cloud Precipice
Patapsco River, Maryland

A high plateau stone with a fairly uniform oval shape.

Base: Walnut, Brian McCarthy ı Collector: Brian McCarthy ı Owner: Brian McCarthy
8 x 6 x 4.5 inches (20.3 x 15.2 x 11.4 cm)

131 ꞁ Unnamed
Northern shore of Georgian Bay, Lake Huron, Ontario, Canada

This dynamic piece of layered quartzite and schist with its multiple fragile fins is suggestive of flow and movement among other things.

Base: Cherry, Tony Ankowicz ꞁ Collector: Tony Ankowicz ꞁ Owner: Tony Ankowicz
12 x 22 x 20 inches (30.5 x 55.9 x 50.8 cm)

132 (right) ı Unnamed
Northern California River

This low rounded stone with its pattern of stripes is suggestive of a crouching tiger.

Base: Wood slab ı Collector: Don Kruger ı Owner: Don and Chung Kruger
8.5 x 5 x 3.5 inches (21.6 x 12.7 x 8.9 cm)

133 ı Wyvern
Northern shore of Georgian Bay, Lake Huron, Ontario, Canada

This highly irregular shaped stone comes to life as this legendary creature.

Base: Cocobolo, Tony Ankowicz ı Collector: Tony Ankowicz ı Owner: Tony Ankowicz
12 x 14 x 8 inches (30.5 x 35.6 x 20.3 cm)

134 | Half Dome
San Bernardino County, California

This stone depicts a mountain range with a cliff face similar to the famous Half Dome in Yosemite National Park.

Base: Walnut, Jerry Braswell I Collector: Hans Thern I Owner: Alex Loughry
5.7 x 13.4 x 6.2 inches (14.5 x 34 x 15.7 cm)

135 | Unnamed
Pacific Northwestern coast

This cobble has a surface pattern resembling a turtle. It is beautifully displayed with a piece of North American driftwood.

Base: Wood slab with driftwood | Collector: Don Kruger | Owner: Don and Chung Kruger
12 x 9 x 3 inches (30.5 x 22.9 x 7.6 cm)

136 ι **Kit Carson's Santa Fe Trail**
Central Rocky Mountains, Colorado

This granite cobble with pink quartz inclusions make an attractive pattern stone resembling a galloping horse.

Base: Maple, Joshua Adamo ι Collector: Nan Morgan ι Owner: Nan Morgan
5 x 3 x 8 inches (12.7 x 7.6 x 20.3 cm)

137 (right) | Cloudy Peak

Eel River Watershed, Northern California

Clouds just below the peak complete the scene of a natural, well-proportioned mountain stone.

Base: Painted wood, Al Nelson | Collector: Ken McLeod | Owner: Paul Schmidt, Jr.
6 x 12 x 7 inches (15.2 x 30.5 x 17.8 cm)

138 | Shoshone Mountain

Wyoming

This reddish mountain-shape stone with its yellow patches and black streaks may appear as an abstract painting. A single base-line cut was made to this stone.

Base: Walnut, Jerry Braswell | Collector: unknown | Owner: Jerry Braswell
6 x 9 x 4 inches (15.2 x 22.9 x 10.2 cm)

139 ı Secluded Sierra Peaks
Eel River Watershed, Northern California

A charming two-peaked mountain stone with a landslide area just left of the tallest peak.

Base: Cherry, Al Nelson ı Collector: Ken McLeod ı Owner: Paul Schmidt, Jr.
5 x 4 x 6 inches (12.7 x 10.2 x 15.2 cm)

140 (right) ⏐ **One Verse Sagan's Song**
Northern Alaska Range, Nenana River, Alaska

This cobble can be viewed as a celestial pattern stone with the Milky Way prominently featured.

Base: Bronze tray, artist unknown ⏐ Collector: Nan Morgan ⏐ Owner: Nan Morgan
8 x 4 x 9 inches (20.3 x 10.2 x 22.9 cm)

141 ⏐ **Rest Beneath This Tree**
Central Rocky Mountains, Garfield County, Colorado

This small glacial moraine, granitic boulder is a wonderful pattern stone that becomes an invitation to sit beneath a spring canopy of blossoms.

Base: Cherry, Nan Morgan ⏐ Collector: Nan Morgan ⏐ Owner: Nan Morgan
8 x 1.5 x 12 inches (20.3 x 3.8 x 30.5 cm)

142 ı Brazil
Pennsylvania

The silky black plateau stone creates a dramatic setting with its overhang and furrows.

Base: Rosewood, Sean Smith ı Collector: Ralph Bischof ı Owner: Ralph Bischof
5 x 8 x 6 inches (12.7 x 20.3 x 15.2 cm)

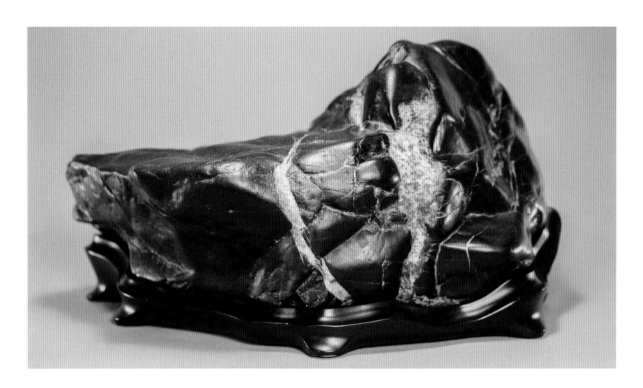

143 | Kimigayo
Jade Cove, California

This natural piece of botryoidal jade is well matched with this antique tray that was originally intended for Japanese flower display.

Base: Bronze tray | Collector: Jim and Alice Greaves | Owner: American Viewing Stone Resource Center
7 x 10.5 x 5.3 inches (17.8 x 26.7 x 13.3 cm)

144 (left) | Unnamed

Eel River watershed, Northern California

This black plateau stone with a waterfall exhibits dramatic color and excellent proportions.

Base: Painted wood, Cliff Johnson | Collector: Ken McLeod | Owner: Paul Schmidt, Jr.
9 x 7 x 10 inches (22.9 x 17.8 x 25.4 cm)

145 | Lost Horizon

North Fork of the Kern River, California

This multi-textured plateau stone is unusual in that it is highly rounded with the mountain centrally located.

Base: Wood, Jim Greaves | Collector: Jim Greaves | Owner: American Viewing Stone Resource Center
4.7 x 6.1 x 3.1 inches (11.9 x 15.6 x 7.9 cm)

146 । California Chrysanthemum Stone

Eel River, Northern California

This natural river-tumbled stone can be displayed in several different orientations.

Base: Pillow and wood, Jim Greaves । Collector: Alice Kikue Greaves । Owner: American Viewing Stone Resource Center
8 x 6 x 3.5 inches (20.3 x 15.2 x 8.9 cm)

147 | Canyon Lands
Vermont

This granitic stone presents a grand landscape with multiple canyons and mesas.

Base: Rosewood, Kathi Maisano I Collector: Ralph Bischof I Owner: Ralph Bischof
2 x 11 x 7 inches (5.1 x 28 x 17.8 cm)

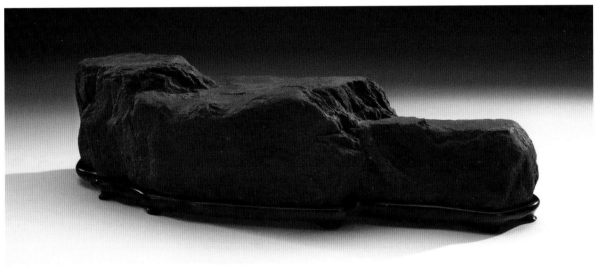

148 | Highlands
New Jersey

This stone of three plateaus shows pleasing proportions and richly detailed steps.

Base: Rosewood, Sean Smith I Collector: Ralph Bischof I Owner: Ralph Bischof
3 x 11x 3 inches (7.6 x 28 x 7.6 cm)

149 | Island Stone
Eel River, Northern California

The lighter green patches on this long, low island-shaped stone may remind viewers of green vegetation on mountain slopes.

Base: Bronze tray, artist unknown | Collector: Jim Greaves | Owner: American Viewing Stone Resource Center
2.7 x 16.2 x 7.5 inches (6.8 x 41.1 x 19 cm)

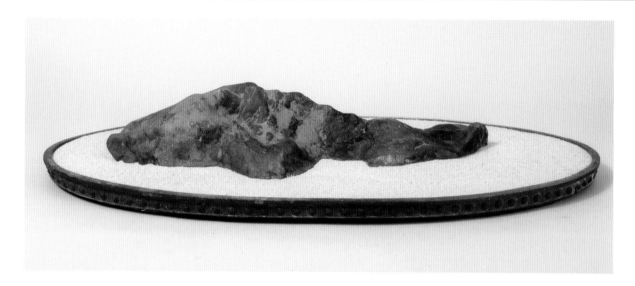

150 | Into The Distance
New Jersey

This understated long plateau stone evokes a mountain range fading into the horizon.

Base: Rosewood, Sean Smith | Collector: Ralph Bischof | Owner: Ralph Bischof
4 x 16 x 5 inches (10.2 x 41 x 12.7 cm)

151 | Tiger-stripe Pattern Stone
British Columbia

A photograph of this North American stone was published in the *International Bonsai Digest* bicentennial edition in 1976. This base replaces the original one that was lost.

Base: Jim Greaves I Collector: Jim Greaves I Owner: American Viewing Stone Resource Center
8 x 7 x 3.1 inches (20.3 x 17.8 x 7.9 cm)

Floating World Editions publishes books that contribute to a deeper understanding of Asian cultures. Editorial supervision and copy editing: Ray Furse, Stephen Elias, and Mary Taylor Jensen. Book and cover design: Liz Trovato. Production supervision: Jane Walsh, Joyce Gu, and Ivan Chan. Printing and binding: Toppan Leefung Printing Limited, Hong Kong. The typefaces used are DIN and Goudy Sans.